VoIP AND UNIFIED
COMMUNICATIONS

VoIP AND UNIFIED COMMUNICATIONS

Internet Telephony and the Future Voice Network

William A. Flanagan

A JOHN WILEY & SONS, INC., PUBLICATION

Published by John Wiley & Sons, Inc., Hoboken, New Jersey
Published simultaneously in Canada

For general information on our other products and services or for technical support, please contact our Customer Care Department within the United States at (800) 762-2974, outside the United States at (317) 572-3993 or fax (317) 572-4002.

Wiley also publishes its books in a variety of electronic formats. Some content that appears in print may not be available in electronic formats. For more information about Wiley products, visit our web site at www.wiley.com.

Library of Congress Cataloging-in-Publication Data:

Flanagan, William A.
    VoIP and unified communications: Internet telephony and the future voice network /
William A. Flanagan
        p. ; cm.
    Includes bibliographical references and index.
    ISBN 978-1-118-01921-4 (cloth)

Printed in the United States of America

10 9 8 7 6 5 4 3 2 1

Dedicated to
My Wife
and Children

CONTENTS

PREFACE

This book intends to prepare you to define Unified Communications (UC) for yourself and then get it to work for you.

Each vendor pulls together from its available products a package of features related to voice, data, messaging, and image communications. That's UC for one vendor, but it's unlikely to match exactly the UC from another vendor. You need a detailed specification to know what you'll see installed.

Second, UC isn't a magic button that solves every problem. On the contrary, careless attempts at UC can create expensive disruptions to your business. Be certain when you deploy UC that it actually enhances your business model or improves processes. Don't do it just because everybody else is doing it. Planning for UC is an ideal opportunity to examine how you work, with a goal of reducing complexity.

Third, VoIP and UC may reduce overall costs in the long term, but it's not free. Nemertes Research interviewed hundreds of companies that deployed VoIP to find the average first-year expense; it was over $1400 per employee. UC features would be additional.

So here's the catch: you can't plan UC very well unless you know what components and functions are available, how they work, how they work together, and how you can use them profitably in your own situation. In addition, some features, for example voice telephony and high-definition video conferencing, will impact your IP network in ways you might not expect. Other features, like Presence, may not operate well across services and vendors. Any number of new features could offer ways to change your procedures that will require retraining of staff.

Defining what you want involves some preparation on your part to learn the basics of the technology, including the vocabulary, so you can speak with some authority. Hence the need for some explanation of what is available, some background for context, and how to use it—my purpose here. To that end, I reviewed more than 7000 pages of published standards, plus data sheets, websites, white papers, and webinars—to save you much of that effort.

My hope is that the practice of VoIP and UC will avoid the complexity that ISDN had to deal with in the United States. In Europe the carriers almost eliminated customer options for the basic rate ISDN service (BRI) used in homes and small offices (two voice channels on a copper pair). An ISDN phone was plugged in and worked. In the United States, the same BRI had (has?)

about 50 configurable parameters, almost all of which are incomprehensible to consumers (and most telco employees).

With some planning and a good deal of luck, a few standard UC profiles will emerge. In place of a long list of parameters to set up a session, the invitation message will carry one description header, "PROFILE = " with a choice of very few values, say something like "StdPhone," "G3fax," or "VideoConf." The savings in bandwidth, processing latency, and equipment design could be huge. The SIP Connect agreement (version 1.1 issued in 2011 by the SIP Forum) is a good start at one area, SIP trunking. Ask your prospective vendors what they are doing to establish profiles and simplify configuration.

As markets mature and users grow more familiar with what's available and what they really want for daily use, the package known as UC will become better defined. Until then, you must specify what you need and want, then ask vendors to bid on your specific UC.

Unfortunately for planning purposes, the market for UC products and services changes daily. You'll have to pick what's best for you from what's available at the time. A book can't offer the very latest product information—that's what the web does. What this book intends to do is give you an overview of typical products and services, with the basis for judging what you find on the web. From that you can hold up your side of the conversation when speaking with sales and technical people. With a clear understanding, you also should be able to respond effectively to the questions and concerns of top management.

Nevertheless, there is hope that the information here will be of great value to those sales people, support engineers, and even newcomers to the industry who want to learn about or clarify their understanding of UC and VoIP. The technical level of the text is designed to include all readers. For those who have been in telephony, there are many references and comparisons to legacy phone services and how UC functions replace them.

To avoid jumping around in the book to understand one concept, some context information is repeated where necessary. Some repetition is not a bug, it's a feature.

At several points in this book you will see warnings and cautions about potential problems and threats to UC services. These statements shouldn't raise undue alarm or create doubt about the migration to UC, but make you aware of issues that telephony managers haven't faced before. For example, there are:

- New and changing legal requirements related to E911 location reporting.
- Hacking threats from Internet connectivity.
- Increased demands on IP networks for high availability and "no-downtime" servers.
- Taxes on Internet services that used to be exempt.
- Large demands for bandwidth from video-and file-sharing applications.

We live in interesting times. I hope this book prevents at least some of your headaches.

—WILLIAM A. FLANAGAN

ACKNOWLEDGMENTS

The work of the Internet Engineering Task Force in publishing the Requests for Comment (RFCs) and Internet Standards helped make this book possible. Only the RFCs, IEEE and EIA standards, ITU Recommendations, and various implementation agreements fully describe the procedures, conditions, exceptions, and options for VoIP and UC protocols. After 10 years as a member of a Technical Committee for the Frame Relay Forum, I have a deep appreciation for the work involved.

Portions of standards appear here where their examples or statements say it best.

The archive of messages from the SIP Forum discussion list provided valuable insights into the practical matters faced by implementers. Thanks to all who shared their knowledge there.

Special thanks and appreciation to the many firms which provided briefings and answered my often detailed questions about VoIP and UC. In random order:

- Smoothstone
- Mitel Networks
- Encore Networks
- Avaya
- Siemens Enterprise Networks
- Cisco Systems
- Juniper Networks
- Broadvox
- OpenText
- Sprint
- Alcatel-Lucent
- Dialogic
- NEC
- Tone Software
- Secure Logix
- Ingate Systems
- Apparent Networks
- ERF Wireless
- AudioCodes

and all the companies mentioned in the chapter examples.

W. A. F.

1

IP TECHNOLOGY DISRUPTS VOICE TELEPHONY

Packet voice, Voice over IP, and Unified Communications (UC) technologies are remaking telephony in a fundamental way that hasn't been seen since the 1960s. Then the Bell System introduced digital transmission and switching inside the carrier infrastructure to replace analog methods. As digital technology spilled over to businesses through the 1980s, a wave of digital PBX's replaced older analog PBXs, key systems, and other forms of analog technology. Today the only remnant of analog in the public switched telephone network (PSTN) is the plain old telephone service (POTS) line, the once-universal service. POTS is being discontinued only gradually, but will probably disappear some day as cell phones, fiber to the home, and voice over cable TV networks continue to replace POTS with Voice over IP (VoIP).

1.1 INTRODUCTION TO THE PUBLIC SWITCHED TELEPHONE NETWORK

Telephones are so simple to use that they hide the complexity inside the network that provides the many features we enjoy. In designing a UC deployment, it's good to understand what UC will replace and extend; that is, what we have used to date.

VoIP and Unified Communications: Internet Telephony and the
Future Voice Network, First Edition. William A. Flanagan
© 2012 John Wiley & Sons, Inc. Published 2012 by John Wiley & Sons, Inc.

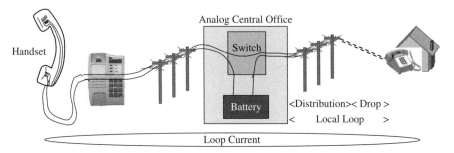

FIGURE 1.1 Current loop from CO battery to phone.

Figure 1.1 describes the original telephone technology, the analog phone or POTS line—Bell's great invention. The phone at the house or office connects to the telco's central office over a 2-wire copper line. The copper wires are twisted to reduce interference from external sources, such as AM radio stations and large electrical motors, but are not shielded by an external metal wrap—hence the term unshielded twisted pair (UTP). Electrical current to operate the phone comes from the battery in the central office; the phone needs no other power supply. Power from the CO was necessary when the first phones were installed because at that time lighting was by gas. Not many homes (and not all offices) had electricity.

Electrical current flows in a loop from end to end, through both phones. The portion of the connection between the customer and the CO came to be called the "local loop." The transmitter in the mouth piece varies the rate of current flow in response to the sound waves from a talker's mouth. Since the current flows in a loop, the same changes occur at the receiver where the miniature audio speaker in the earpiece reproduces the talker's voice.

The system grew more complex as automatic switches took over from live operators, but the legacy signaling system is outside the scope of this work. For more information, see *The Guide to T-1 Networking*.

1.2   THE DIGITAL PSTN

The digital revolution hit the network in the 1960s with the deployment of channel banks. These multiplexers combine 24 analog circuits (2-wire POTS, 4-wire E&M, and other types) onto two twisted pairs, one for each direction, in the digital format that came to be known as T-1.

The reduction in wire count applied first on the trunk lines between central offices. The COs had room to house the new equipment, but more important, the cable ducts buried in the streets of major cities were filling up. The phone company couldn't easily add more copper cables to fill the need for additional trunks between switches.

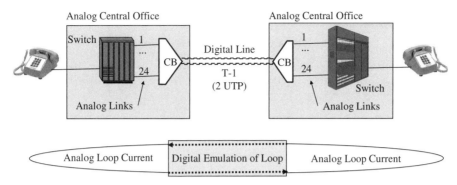

FIGURE 1.2 Channel banks between analog switches.

There was an added benefit to digital transmission: better sound quality. In most situations the 1's and 0's on the T-1 line survived intact, even if some analog noise were added by arcing motors, radio stations, or other sources. The receivers in the channel banks correctly recognized even a slightly distorted "1" as different from a "0," so the output sound wasn't impaired.

Digital transmission between analog switches looks like Figure 1.2. The transition from analog to digital for inter-office trunks was relatively easy and left other network devices in place. In this early form of digital telephony, the capacity of the T-1 line divides into 24 fixed channels based on time division multiplexing (TDM). That is, the 24 analog inputs take turns in strict rotation to send one byte of digitally encoded voice that represents a sample of the analog input loudness (the instantaneous volume level). The receiver converts that byte into a matching output level.

The Nyquist theorem regarding information transmission proved that if the samples were sent at a rate that was at least twice the highest audio frequency of the analog input, then the reproduction in the output at the receiver would be consistent with the input (reproducible results). Design compromises and precedents from analog telephones settled on a voice frequency range of 300 to 3300 Hz. Cutting off everything under 300 Hz eliminated AC hum and matched the limited capability of handset hardware to reproduce low frequencies. The top of 3300 Hz fit within what was then the standard for analog multiplexing: 4000 Hz for each analog channel.

To ensure that the sampling rate exceeded twice the highest voice frequency, the chosen sampling rate was 8000 per second. Each channel, then, generates $8 \times 8000 = 64 \, \text{kbit/s}$. This rate, the lowest in the digital multiplexing hierarchy, is numbered the way engineers start to count, with zero. Digital signal 0 (DS-0) is the fundamental building block of the TDM hierarchy in circuit-switched voice networks.

The T-1 bit rate is the sum of 24 channels plus an extra framing bit per cycle of 24 channels, a T-1 frame (Figure 1.3). This format continues in use as the way bits are organized on a primary rate interface (PRI) ISDN line. One of

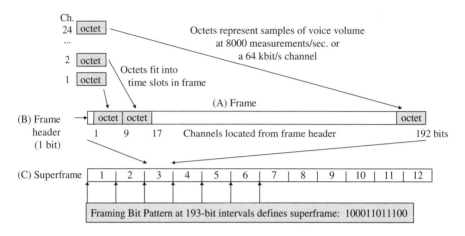

FIGURE 1.3   TDM frames showing the basic concept, a T-1 frame, and a superframe.

the DS-0s on a PRI, the D channel, carries only signaling messages, or what was called data because it wasn't voice.

In any time division multiplexer, the basic frame consists of a string of bits marked in some way by a unique signature element which defines the frame (A). Some link protocols reserve a "start of frame character" that has no other use and never appears inside a frame.

In T-1 and PRI, the marker is a single F bit (B). One bit alone doesn't allow a receiver to identify the start of the frame. The structure of a superframe (C) built up from 12 frames makes room for a fixed pattern across the superframe: 100011011100. The framing bit pattern allows the receiver to identify the locations of the F bits and from them the groups of bits associated with each channel. An extended superframe (ESF) of 24 frames uses a more complex pattern of F bits that includes a data channel.

The result is the now familiar T-1 bit rate:

$$8000 \times [(24 \times 8) + 1] = 1.544 \text{ Mbit/s}, \quad \text{which is a DS-1.}$$

Keep in mind that channel banks operate continuously. For each analog input (even if it is silent) the time slot on the DS-1 formatted line carries a byte of "sound" in every one of the 8000 frames per second. The capacity of the line is dedicated to the port on the channel bank, whether or not it is in use. In effect the digital transmission system of channel banks and T-1 lines (the original digital transmission technology) emulates the current flow in the analog local loop. T-1 transmission could also be compared to a moving sidewalk seen at most major airports. It runs at a constant rate whether or not there are passengers on it.

More precisely, the multiplexing format is DS-0; T-1 is a transmission technology on two twisted pairs that requires a repeater every mile but can be extended up to 50 miles. Digital subscriber line (DSL) equipment has largely displaced T-1 in local loop, with a longer reach at 1.5 Mbit/s without a repeater, but is more difficult to extend. Optical fiber now dominates between COs.

Some references to TDM-defined voice channels call it wasteful of bandwidth, but such a judgment should also take into account two other factors:

1. **Low overhead:** only 1/48 of a bit per octet sample is enough to identify the channels. Only half the F bits are used for ESF framing; the other 12 F bits are a data channel.

2. **Low latency:** each channel has a reserved spot in every frame. The latest byte from the speaker's voice digitizer need wait no more than 1/8000 second (125 microseconds, μs) for the next frame to carry it away on the T-1 line.

Dedicated capacity per call prevents interference between users. One caller shouting can't affect another who is whispering. With digital transmission, quality is consistently high. All callers who get connections receive the same high quality of service. Hold these thoughts for comparison to VoIP later.

Years after the first T-1 lines were installed between central offices, subscriber lines remained individual copper pairs from the switch in the CO to the telephone. Huge cables with thousands of pairs, laid from the CO to a large office building or to a residential neighborhood, had to be spliced by hand each time another reel of cable was added to the run. The biggest reel could hold as little as 1000 ft of a 4000-pair cable. Pieces of cable rarely exceed 1 mile, and the largest cables were installed mostly within large buildings.

The standard service area for a CO is measured by the length of its local loops: 12,000 feet is a common goal for the longest loops out of an office, which typically required splicing those cables once or twice.

When CO switches became digital, the channel bank was adapted to become an extension of the CO switch, with digital T-1 connections for most of the distance to the building or neighborhood. Splicing in the distribution network was reduced by a factor of 12 (or as much as 48, as described below).

In a sense, the original POTS is almost gone because the copper pair from the analog phone no longer reaches to the central office battery that powers the switch. In many areas, particularly those built up in the 1980s or later, the analog line ends within the neighborhood at a remote terminal (or channel bank). You can see the pedestal cabinets that hold them by the side of the road (Figure 1.4). From there the connection to the central office is a digital transmission line on copper or an optical fiber.

The channel bank grew into the subscriber loop carrier (SLC) with up to 96 analog ports. It could be placed in a closet of a building, or into a free-standing cabinet near a cluster of homes. As Figure 1.4 shows, the analog ports on the SLC still power the phones over separate UTP lines. The payoff for the telephone company was a huge reduction in the distribution cabling where T-1 links (and, later, optical fibers) replaced the individual copper pairs.

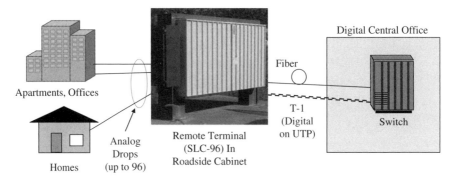

FIGURE 1.4 Pedestal cabinet that holds a remote terminal (SLC-96) for POTS service to a neighborhood.

One negative was the need to power the SLC. Often an electrical utility meter is visible on the cabinet.

Recognizing that not every phone wants to call at the same time, the SLC "oversubscribed" its lines to the CO. In residential areas the 96 analog ports on the SLC often share a single T-1 from the SLC to the CO. The SLC, integrated into the switch's logic, assigns a channel on the T-1 only during an active call. In business environments where more simultaneous calling is common, the phone company will install up to four T-1s if necessary, which allows all phones to call at once. Today a pair of optical fibers can carry all the calls from any number of SLCs at a site. Later sections will compare this circuit-based local loop technology with packet-based links such as SIP trunks.

Oversubscribing at the SLC didn't change much for subscribers. Customers wouldn't notice unless some event triggered mass calling. However, CO switches are also limited in the number of calls that they can set up per minute because the number of modules that receive dialed digits from a phone is much smaller than the number of phones served by the switch. A caller needs one of these modules to place a call, then the module is freed to handle another request while the first call remains active. In the unlikely event you have ever had to wait for dial tone after picking up the handset on a POTS line, you have waited for one of these modules to become free. Call setups per hour is a valid metric for VoIP servers as well.

To summarize the result of the digital revolution, Table 1.1 lists the attributes of phone calls made on circuit-based analog and digital system. Digital PBXs preserved the ability to power phones over the drop cable. Depending on the vendor, the power may have been on a phantom pair (Figure 1.5) or a separate copper pair in the same cable. A phantom pair derives from transformers at each end that couple the audio but keep the dc power on the drop wire.

This phantom pair for power distribution is seen again in IP phones with Power over Ethernet (PoE) as defined in the IEEE standard 802.3af. The digital

TABLE 1.1   Characteristics of phone calls on analog and digital networks

Phone Call Property	Analog Network	Digital Network
Sound quality	Often quite good for local calls; weaker and noisier for long distance calls	Almost always uniformly high
Susceptibility to noise	High originally when transmission and switching were all analog; limited lately as T&S are now digital	Very limited
Distribution cable	Copper unshielded twisted pair	Optical fiber
Drop cable	UTP	UTP
Phone power source	Battery in central office (or SLC)	Local power, fed either from PBX or from wall transformer
Echo	Always a concern; requires fine tuning amplifier gains and line losses (to minimize amplitude) plus echo cancellers	Digital echo cancelers (in media gateways and phones) make echo undetectable on most calls

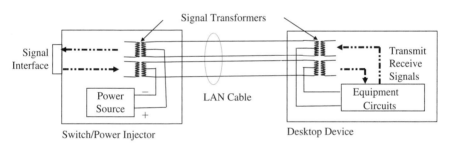

FIGURE 1.5   Phantom power over two twisted pairs.

revolution fifty years ago retained some concepts and features from the analog technology. In particular, digital switches reserved capacity in defined circuits or channels for each call across the switch and over connected transmission lines (Figure 1.6).

To set up a connection between digital trunks, a circuit switch starts a repetitive process that accepts the octet in a time slot on the inbound port, buffers it for a very short interval, and places it in the appropriate time slot in the next frame leaving the outbound port. The process works symmetrically, 8000 times per second, transferring octets in both directions between the connected time slots. Such a switch is also known as a time slot interchanger (TSI). The transfer delay averages about two frame times or 250 μs. SLCs

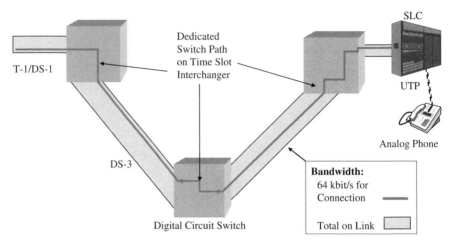

FIGURE 1.6 A circuit-switched connection occupies dedicated capacity in switches and transmission lines for the duration of the call.

behave similarly, dedicating a TDM channel from the SLC to the CO for each call on an analog port.

The channel exists end to end only for the duration of the call. A call clears when the TSI mapping from input to output disappears and the trunk time slots become available for new assignments.

1.3   THE PACKET REVOLUTION IN TELEPHONY

The packet revolution changes the network fundamentally, yet some elements are very similar.

Since human speech is analog, voice on a digital IP or packet network must be converted to a digital format, an encoding process that may be identical to that in a channel bank or a digital circuit-switched network. That is, packet voice often is encoded as pulse code modulation (PCM) as defined in G.711 for the original channel bank. But the bytes of data no longer stream immediately and at a constant rate over a dedicated 64 kbit/s channel.

In voice over IP (VoIP), the digital information is saved up for a short interval (typically 10 or 20 ms), then put into a packet and sent in a burst over the digital line at the line's bit rate, usually much higher than 64 kbit/s such as Ethernet at 100 Mbit/s.

Where a T-1 transmission is a moving sidewalk, packet transmission is more like a high-speed shuttle train between terminals. Each car takes on a number of pedestrians (digital bytes) over the time in a station and moves them together and at higher speed. Both the trains and moving sidewalks could have the same capacity, able to carry the same number of passengers per hour (octets per second). For either transport method, the operations at the ends (buying tickets and going through security, or encoding and playback) deal with one individual/byte at a time.

Don't rely too much on the metaphor. Keep in mind that voice channels contain flows of information bytes, not individuals. A moving sidewalk accepts any mix of people, whereas a T-1 frame dedicates each byte position to a specific channel. A shuttle train accepts random groups of individuals, whereas a VoIP packet represents the information of only one conversation. The concept of a stream is the flow of packets or bytes related to a single function or conversation.

1.3.1 Summary of Packet Switching

Because many packet connections can share one line, each packet must carry its destination address so that the network knows where to deliver it. To mix a metaphor, each train must be routed to the proper terminal, or the destination is put on the front of the bus. The addresses take several forms, depending on how they are used by the network. Addresses plus additional control information constitute the headers on a packet.

To ensure a common understanding of terms for this book, this section will describe how packet networks operate. Figure 1.7 shows the headers that make up a typical voice payload packet. A more detailed discussion appears later.

For this and other descriptions of packets, the convention here is that bits are transmitted as if the diagram reads like English text; that is, from left to right starting in the top row and then the next row below until the end of the packet. Within an octet, the least significant bit (LSB) is sent first. Header diagrams are upside-down compared to the standard representation of a protocol stack.

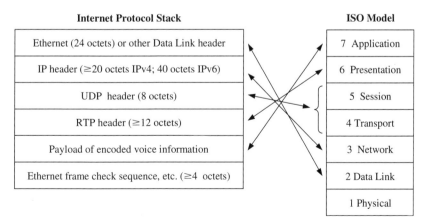

FIGURE 1.7 Internet protocol headers on a VoIP packet roughly corresponding to layers of the ISO model of a protocol stack.

The International Standards Organization (ISO) diagram shows seven layers. The bottom is the physical layer 1: copper, optical fiber, radio, or the string between two tin cans. Protocols occupy layers 2 through 7. The ISO data link, layer 2, is very close to the Internet data link and may use the same protocols such as Ethernet, frame relay, and generic encapsulation protocol. In the legacy data environment there are many more layer 2 protocols not of concern to this discussion of VoIP and UC.

While L2 is at the bottom of the ISO model, the header for the L2 protocol appears at the top of the packet header diagram. It is sent first because it goes the shortest distance—only to the other end of a transmission link.

The L3 ISO protocol for the network connection comes next. This is the position of the Internet Protocol, IP, whose function is to send packets to another host or hosts which can be anywhere on the Internet. An IP header can take a large number of hops from device to device as the packet finds its way across the network. IP has two main characteristics:

- IP works on a best-effort basis, with no guarantees of delivery.
- IP is connectionless. The network accepts IP packets at any time—the network does not require any preparation to receive a packet for a new address.

This kind of service is also known as a datagram service.

IP doesn't guarantee delivery of information; this is a function of the next protocol at the ISO transport layer (L4), which can guarantee delivery of packets and in the proper order. On the Internet, Transmission Control Protocol (TCP) most often performs this function. TCP uses sequence numbers to spot missing packets and ensure delivery order. Error checks recognize transmission or bit errors. The sending TCP process saves packets until the receiver acknowledges receipt, in case a packet must be resent to correct an error. For voice packets, the User Data Protocol (UDP) occupies L4 and L5, so there is no real ISO transport layer error correction in the case of VoIP.

A host that receives a packet needs to know what to do with it—which process or application should deal with it. The ISO layer 5 protocol establishes a session between applications; that is, it identifies a sequence of packets associated with one process or transaction. The port numbers in TCP and UDP headers identify the associated process at each end.

ISO protocol layers are very specific to their functions, with defined interfaces between them. The idea is to allow changes at one layer without affecting any other layers, above or below, because the application program interfaces (APIs) are constant. The Internet protocol stack doesn't line up exactly with ISO, but the goal of interchangeability of elements is the same. Users can deploy a hardware improvement or an updated portion of software without disruption to items on other layers. The interfaces between layers remain constant or change very slowly. The adoption of IPv6 would be much more difficult if IP were not confined to L3.

The presentation layer 6 is not often seen separately from an application. That is, the author of an application usually decides how it will appear to users. There are libraries of software functions that present information graphically, or enhance text displays. For VoIP, the Real-time Transmission Protocol (RTP) operates above ISO layers 2, 3, and 4 to provide functions tailored to voice and video applications. RTP is not strictly presentation, and not the full application, but provides what's needed to support voice and video transmission—or, any streaming medium.

"Applications" are what most users think of as software, rather than layer 7. References to layer 7 are often meant to include any application.

1.3.2 Link Capacity: TDM versus Packets

There are two schools that put entirely different emPHAsis on the sylABles defining bandwidth efficiency. The outcome of the discussion impacts what call capacity a network designer will attribute to a link.

The advocates for "everything over IP" point out that channels defined on a transmission line get in the way of allocating bandwidth when and as needed. An open pipe T-1, for example, carries every packet at 1.536 Mbit/s, the data capacity after deducting the 8000 framing bits per second from the line bit rate of 1.544 Mbit/s. A channelized T-1, such as those used as voice trunks between a central office and a PBX, carries each channel at only 64 kbit/s. A packet transmitted on a DS-0 channel takes 24 times as long to finish as a packet sent on an unchannelized T-1.

Traditionalists point out another way to measure efficiency: the ratio of information bits to total bits on a link. For channelized voice traffic a full 24 channels represents 1.536 Mbit/s of voice and signaling information out of 1.544 Mbit/s, or about 99.5%.

What really matters is how many conversations will that T-1 access link support at one time. In a legacy TDM format the answer is 24. When the mode is VoIP, the answer varies over a wide range.

Packets require headers in addition to the information bits. Compared to a TDM channel, the number of bits per second for a conversation is higher if PCM voice encoding is packetized. That is, chopping a 64 kbit/s voice signal into packets requires adding 44 or more bytes (can exceed 64 bytes) to each 20 ms block of voice information. That's 44 to 64 bytes added per 128 bytes.

For the simple use case of PCM and IPv4 on an Ethernet link, 64 bytes of header on 128 bytes of voice information raises the bandwidth needed in each direction to 96 kbit/s. Additional bandwidth is needed for the idle intervals required between packets on some Ethernet interfaces, optional headers on IP packets that belong to a virtual private network (VPN), and additional traffic to support authentication and other functions.

Common practice allocates at least 80 kbit/s of bandwidth for each voice channel encoded with standard PCM. To include all packet headers, it is more realistic to assume 100 or 180 kbit/s per conversation for link capacity planning. The effective number can vary when the system applies various methods to save bandwidth, described below. For one, compressing the voice information to 8 kbit/s (e.g., with the G.729 algorithm) doesn't reduce the headers, so the bandwidth per channel for link sizing drops to around 50 kbit/s.

A major consulting firm reported that a T-1 line could support 50 conversations using VoIP, more than double the TDM capacity. To reach that density requires additional processing.

Header compression reduces the bandwidth per voice conversation. Since the headers are pretty much the same in packet after packet (addresses are constant, sequence numbers and time stamps increment predictably), it is possible to substitute a "token" value to represent the full set of headers. Several RFCs define the process, in which the sender substitutes 1 to 4 bytes for the complete 44+ bytes in the original headers, not including the data link protocol. In this form of compression there are other headers that aren't compressed, for example, an Ethernet, Frame Relay, or Multi-Protocol Label Switching (MPLS) tag to multiplex connections on a link.

Keeping with the simple use case, and adding the minimum Ethernet overhead (24 bytes) to a compressed voice payload (16 bytes) produces a total packet length of 44 bytes. The headers repeat 50 times per second, requiring 17.6 kbit/s. Replacing Ethernet with a data link protocol that uses a much shorter header, like Frame Relay or HDLC, can reduce the full-duplex bandwidth per conversation to about 12 kbit/s. More than 50 of them will fit on a T-1.

Carriers often use double MPLS headers (Figure 1.8) to simplify their internal configurations, but those headers don't require bandwidth on access

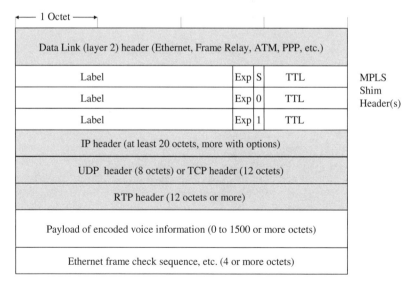

FIGURE 1.8 MPLS labels add to header size but simplify packet forwarding and support traffic engineering for voice quality.

lines, only on the carrier's backbone. MPLS enables a network to set up static routes in tables (like those shown later in Figure 4.1), to ensure voice packets follow a physical path that introduces minimum latency.

On the wide area network (WAN) and the fastest Ethernet links (full duplex connections with separate paths for each direction), the transmission equipment can queue packets and launch them head to tail with only a short separator between. On a local area network (LAN) based on a slower Ethernet, each packet starts with a preamble of a bit pattern that lets other hosts know a packet is coming and at what bit rate. That interval takes bandwidth too.

The most significant block to high transmission efficiency in packet networks is the problem of congestion handling. Switches and routers store packets they can't send immediately in a local memory buffer. When that buffer fills, the only available relief is to discard packets.

Discards work well with data connections based on Transmission Control Protocol (TCP) because TCP client and server software in hosts recognizes lost packets as congestion and slows the transmission rate. Reduced throughput isn't acceptable for voice, which depends on a constant-rate stream of information. Traditionally voice has been a constant bit rate service (64 kbit/s) with no speed variations.

VoIP operates on User Datagram Protocol (UDP), which has no mechanism to slow transmission. Variable bit rate compression algorithms exist, but typically they are based on the complexity of the talker's voice rather than network congestion. So to avoid dropped VoIP packets, the best practice is to allocate no more than 40 to 60% of a link's bit rate to voice service. The rest can be used by TCP connections, if the routers and switches prioritize voice and discard only data packets.

For comparison, the DS-0 channel of 64 kbit/s operates with minimal latency, at full capacity, in dedicated bandwidth for each call. No channel suffers from congestion after it is connected—degradation in service consists of the busy signal and blocked call attempts.

Without call admission controls on VoIP systems, new voice connections can overload a link and degrade the perceived quality for all users. Table 1.2 summarizes this and other differences.

1.3.3   VoIP and "The Cloud"

This book intends to describe in considerable detail how VoIP and UC work. The components of a VoIP/UC system work the same regardless of where they are located physically. The phones will be on desktops, and the network links, routers, and switches will be where needed to connect everything. But the servers, databases, and security appliances can be anywhere on the Internet or a private network. Enterprises will put some servers on premises and others in outsourced data centers or carrier central offices. Hosted VoIP puts the call-processing power at the vendor's site.

IN SHORT: Reading Network Drawings

Figure 1.9 shows a conventional way to represent a LAN. For readers coming from a voice background, the single horizontal line derives from early Ethernet, which consisted of one coaxial cable connecting all the local terminals. There was no switch, only the passive cable and attachment units.

Coax contains a center copper conductor surrounded by insulation and then a braided wire tube-like covering that shields the inner conductor from electrical interference and acts as the second conductor. Ethernet attachments for all the local terminals penetrated the braided shield to connect to the center wire of the coax. Sharing a wire, all the terminals on the LAN received all the bits transmitted by every host. The drawing replicates this topology of a core with branches.

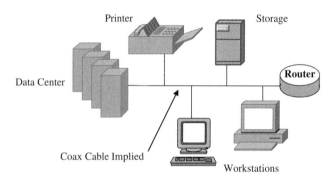

FIGURE 1.9 Depiction of LAN connections derived from early Ethernet.

Is that a cloud? Perhaps, but it may be better to separate two concepts whose overlap can confuse:

- *Hosted* service places the servers in a vendor's site.
- *Cloud* service implies more about the vendor's infrastructure. It is virtualized and clustered to maximize availability (uptime), flexibility (standing up applications quickly), and expandability (adding more processing power, memory, or storage as needed).

A more realistic topology today is for each terminal to connect to its own port on an Ethernet switch. This star-shaped layout is harder to draw accurately. Logically, it requires that the switch be shown and not assumed. Whenever a drawing like Figure 1.9 appears, you can safely assume it really looks like Figure 1.10.

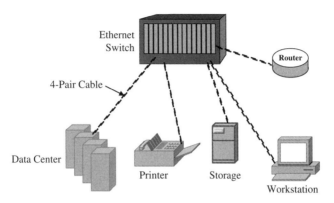

FIGURE 1.10 Actual topology of Ethernet LAN; each terminal has its own switch port.

Network management or automapping software that probes the network for devices typically produces this second format of drawing to represent the network. The additional detail is needed to include every device and to allow each port on each device to carry a label for its IP address, VPN assignment, user name, and so forth.

Private cloud infrastructure may be ideal for your own data center. If you outsource VoIP, cloud infrastructure is a good feature to look for, but not a guarantee of 100% uptime. But for the purposes here, "cloud" is not integral to VoIP; it is a feature of the hosting provider, which could be you. Be sure the VoIP/UC feature set you want is available from a "cloud vendor" and that a cloud is what you need.

TABLE 1.2 Differences between TDM and packet telephony

Feature	TDM Digital Telephony	Packet Voice over IP
Information flow	Constant bit rate in channel	Bursts at line rate separated by pauses
Switching technology	Bytes moved from input channel to output channel at constant rate by TSI	Packets forwarded from input port/line to output port/line as possible
Connection resources in switches and transmission	Dedicated to each call for its full duration	Calls share switch queues and transmission paths; call occupies resources only while packet moves
Telephone number (address)	Assigned by carrier, permanent, fixed location unless ported (not including cellular)	User's network address, from ISP or self-assigned; can be a URL similar to email address, which is usable anywhere
Latency (delay)	Fixed, minimized, not affected by other calls	Can vary, depending on path and congestion from other calls
Echo	Canceled by carrier	System may not create echo, but will cancel if there is a potential to create it
Security	Strong; proprietary software on purpose-built hardware running over dedicated cables	Similar to data network; may be exposed to Internet, voice usually shares cables with data
Equipment costs	Significant; often very high for add-on functions like voice mail, auto attendant, IVR	IP phones more expensive than basic analog, comparable to digital phones; call control software migrating to commercial servers, reducing hardware costs; voice mail, auto attendant, conferencing often included
Operating costs	Move/add/change requires service call; expansion may be expensive for new modules	End user may move phone, which can register its new location automatically or with user authentication.
Maintenance	Typically very little; replace back up batteries, clean fans, possibly update software in 2 to 5 years	Similar to data network; software patches for servers, applications; scan for viruses; replace hard drives; may need new test equipment for voice
Product life expectancy	10 to 20 years; cooling fans may be only moving parts (no hard drives, except for voice mail)	Similar to data network gear, as little as three years before software updates force some telephone replacements (servers should last > 5 years)

2

TRADITIONAL TELEPHONES STILL SET EXPECTATIONS

The analog telephone was a mature product more than 100 years ago. The POTS residential line of today is what every line was in those days. Amazingly those old phones will work on today's analog service. Many central office switches still accept rotary (pulse) dialing as well as dual-tone multifrequency (DTMF or TouchTone) signals. For longevity and preserved backward compatibility, it's hard to find anything as durable as POTS.

Telephones are so ingrained in our lives that many users are not aware of how the traditional service sets our expectations in many ways. There's a story, allegedly true, about a call to a computer support desk from a person whose computer wouldn't turn on. The support agent asked the caller to look at the back of the computer for loose cables. The caller said that was impossible because it was too dark—the power was out.

The expectation for phone service (powered from the CO) was transferred to a computer (on local power). It's hard to see how old telephone expectations won't continue to apply in some significant ways to telephones and telephone service after the packet revolution.

VoIP and Unified Communications: Internet Telephony and the Future Voice Network, First Edition. William A. Flanagan
© 2012 John Wiley & Sons, Inc. Published 2012 by John Wiley & Sons, Inc.

2.1 AVAILABILITY: HOW THE BELL SYSTEM ENSURED SERVICE

As described in Chapter 1, analog phones draw operating power over the local loop from the telco. To ensure continuity of operation, COs contain large banks of batteries, typically enough for over 8 hours of operation, powering the switch as well as the phones. Many COs have standby generators to sustain service indefinitely by taking over from the batteries during extended outages. The tradition was that the phones worked unless the CO burned down (which happens, if rarely).

The Bell System standard for the reliability of any piece of equipment, or the service provided by a set of similar equipment, was 99.999% uptime. That's just over 5 minutes of outage per year (Table 2.1).

When there are multiple devices with that level of reliability, the overall service can have a slightly lower uptime target. The overall availability of a string of elements in series is approximately the product of the individual availability figures multiplied together. When redundant devices back up each other, their availability increases.

Networks to support business services used to be designed to approach "five-nines" overall, not just for each device. Residential service wasn't quite as robust in practice—one carrier's Fiber to the Home (FTTH) service advertises only "99.9% reliability." Residential tariffs also had looser time frames for repairs compared to business services (good reasons that business lines cost more than residential). Overall, however, everyone expected the phones to work.

Cellular phone customers, while trading off reliability for mobility, still harbor some of the old expectations. Carriers recognize that attitude by advertising "most reliable network" and "fewer dropped calls." Still, in the back of the big-box store under a metal roof, cellular service may not be available at all. Loss of signal in that limited situation might be acceptable, but loss of service on a desktop business phone won't be tolerated.

Bottom line: a VoIP implementation must be designed for high availability.

TABLE 2.1   Uptime percentage versus downtime

Uptime, %	Downtime/Year	Downtime/Month	Typical Application
95	>18 days	36 hours	Not acceptable anywhere
99	>3.6 days	7 hours	
99.5	>1.8 days	3.6 hours	
99.9	8.7 hours	43 minutes	Advertised for residential FTTH
99.99	53 minutes	4.3 minutes	
99.999	>5 minutes	26 seconds	PSTN business service
99.9999	31 seconds	<3 seconds	Core network device

Note: Downtime may occur at any time of the day and need not be continuous. For example, 99% uptime allows service outages for 12 periods of 7 hours, all of which could be full business days.

Keeping phone service highly available after converting to an IP network requires some changes in attitude and expectations among traditional data operations people who run the IP network. Increasingly business and financial transaction services on data networks also demand high availability, so the network infrastructure and the people who maintain it have moved away from practices such as:

- Taking down a service for hours at a time to upgrade hardware or to patch software.
- Rebooting a server or router "to see if that clears the problem."
- Accepting delays of hours or days to restore a service after a hardware failure.
- Allowing a local power outage to interrupt services.

Technology advances linked to the concept of "cloud computing" are a big help to availability. Server clusters that provide redundancy and load sharing prevent a single hardware or software failure from halting a service. Virtualization goes further, allowing a new server to come on line automatically when demand increases or a working server drops off after a hardware failure. The ability to move an application from one server to another—without service interruption—also means that managers need not scheduled down time to replace, upgrade, or expand servers; install patches; or upgrade software. The optimum process is to add a new server to the cluster, then shift traffic away from the server to go down until it is idle. That should be doable during normal business hours, not overnight or on a holiday.

2.2 CALL COMPLETION

The "fast-busy" signal has been a rarity for decades. The percentage of calls that fail to complete is so small that many users are not aware of a problem in this regard on the PSTN—because there isn't one. Trunk capacity among central offices is almost always ample to accept all calls placed. The occasion of an extreme emergency will be an exception.

The circuit-switched PSTN has hard limits on the call capacity of each line, from POTS (1) to a fiber (thousands). When the DS-0s between COs are all in use, it is not possible to add another call, and the response is the fast-busy signal, which indicates a lack of resources in the network.

Deploying VoIP for long-distance trunks has expanded capacity and increased flexibility in routing calls, so there is enough capacity almost all the time. On a private network, legacy or VoIP, there is a greater possibility of resource limitations.

Pieces to consider when evaluating or designing an IP network for VoIP are as follows:

- **Call processor capacity:** some server software has a limit on the number of simultaneous calls. Hardware limits need planning; for example, memory to hold the states of connections.
- **Media gateways:** similar to planning for legacy PBXs, the number of TDM trunks beyond the gateway is a strict limit.
- **IP trunks:** unless limited by a H.323 gatekeeper or similar agent, an IP line may accept more calls than it can carry reasonably. The quality of all connections on that link may suffer.
- **SIP trunking:** the carrier supplying the service may impose a limit on the number of calls at a time that is unrelated to the bandwidth of the access link or capacity of servers—you get what you pay for.

2.3 SOUND QUALITY: ENCODING FOR RECOGNIZABLE VOICES

The design of the PCM encoding scheme in channel banks was very clever. It had to be, given the state of electronics in 1960. A key requirement was that voices be recognizable, even at low sound levels, when people are whispering, yet the phone had to reproduce loud shouting. The solution was companding: compressing/expanding. The algorithm is defined in ITU Recommendation G.711, which often lends its number to pulse code modulation (PCM).

Selection of the design parameters (easy in hindsight) could have gone something like this:

- Carry over into the digital design the basic 4 kHz voice channel audio bandwidth, previously proved in analog multiplexing practice and the design of analog handsets.
- Apply the Nyquist theorem, which says that to ensure an accurate digital reproduction, the sampling rate must be at least twice the highest analog audio frequency (sampling has to catch every zero crossing, when the sound pressure changes from positive to negative); $4000 \times 2 = 8000$ per second.
- Measure the voice signal at each sample accurately enough to ensure voice recognition at low sound levels. To capture the nuances required a 16-bit analog-to-digital converter (ADC) with 64 K possible outcomes to each measurement.
- Compact the 65 K possible values from the 16-bit ADC to 8 bits, 256 possible byte values, by using a logarithmic scale and allowing fewer values at louder volumes.
- Establish the DS-0 channel: $(8000/s) \times 8$ bits $= 64,000$ bits/s.
- Assemble 24 DS-0s into a DS-1 with the addition of 8000 framing bits per second.

The cleverness was to selectively map the detailed measurements to only 255 values (all zeros isn't allowed, to preserve electrical pulses, the logical 1's, on the

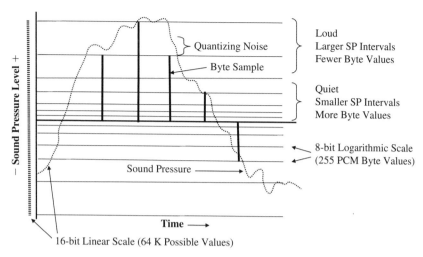

FIGURE 2.1 Companding concentrates voice encoding information at lower sound levels, and reduces the number of bits needed.

transmission line). The full range of sound pressure levels, both positive and negative, then fits into one byte. Some "rounding" is necessary to assign a range of 16-bit measurement possibilities to only one 8-bit result, the byte value. The difference is quantizing noise (QN).

Figure 2.1 shows how this mapping places more of the byte values at low loudness, with fewer values for very loud sounds.

- More possible byte values at lower volume levels preserve the ability to recognize a voice during normal and quiet talking.

- The mapping cuts off very loud input measurements, capping the receiver output level and preventing damage to the network or painful volume levels for the listener.

- Distortion generated in the analog components of a telephone (the microphone, the earphone) increases with high sound levels and masks the distortion of the QN approximations. At loud volumes nobody notices errors introduced by mapping to a limited number of values.

PCM is very mature and inexpensive, making it an easy choice for hardware designers. PCM represents the wave something like a picket fence where the height of each picket represents the volume of the sound as it is sampled at that instant. The tops of the pickets describe the sound wave from the speaker and are reproduced by the receiver. Pickets are 1/8000 of second apart.

There are two variants of PCM. In the United States and Canada where transmission links are based on the T-1 standard of 1.544 Mbit/s, the mapping is called mu-Law, also written with the lowercase Greek letter mu as μ-Law. In areas that base the transmission hierarchy on E-1 at 2.048 Mbit/s, companding is A-Law. The difference arises from how the 16-bit measurements are assigned to the byte values.

The two laws are significantly different (as described in G.711). When configuring equipment you should set both ends to the same law version. The result of a mismatch (the source configured for one, the receiver for the other) is a noticeably distorted but still understandable conversation. While not common, checking for a "law" mismatch should be part of troubleshooting a problem in sound quality.

To express both positive and negative sound pressure, the sound pressure average with no audio input is set to the middle of the range (128). The most negative pressure during the loudest input registers, in decimal format, as a 1 on the scale; 255 is the highest positive value. The most significant binary bit is also called the sign bit, since it is 1 for positive pressure and 0 for negative.

Many packet voice systems use straight PCM encoding. It fulfills expectations for voice quality, is cheap to produce with commercial chips or DSPs, and can pass both DTMF and fax signals (both of which were designed for PCM). Compressing PCM to lower bit rates doesn't reduce the quality by much, as perceived by a human listener. Fax and modems are, in general, another issue, as described elsewhere in this book.

Compression can have an upside. The 64 kbit/s TDM channel, or a packet stream with that payload capacity, can carry a voice signal compressed from a wider audio bandwidth. For example, an ADPCM algorithm requires less than 64 kbits/s to transmit "high-definition" voice with a 7000 Hz audio frequency range, more than double the 3300 Hz of PCM. The higher frequencies reproduce hard consonant sounds better and improve intelligibility.

Rather than compress PCM to save bandwidth, VoIP can apply higher quality encoding to produce better sound. Naturally it's a bit complicated, but well within the capabilities of DSP chips. The method, as described in ITU Recommendation G.722, samples the audio input 16,000 times per second, double the rate for PCM. The ADC outputs 14 bits, leaving two bits per pair of octets to synchronize the operational mode between sender and receiver. Intended for a dial-up DS-0 channel, G.722 can carry an 8 or 16 kbit/s data sub-channel in addition to the voice, but this feature is unlikely to find application in a VoIP system where a separate packet stream for data offers more flexibility and increased capacity.

One way to think of the G.722 process is to split the audio input, with two band-pass filters, into two sub-channels of 0 to 4 kHz and 4 to 8 kHz. Each sub-band is then encoded separately in ADPCM. The two results are packaged into a single DS-0 which can be sent in a circuit-switched channel or packetized like PCM. The overall process is called Sub-Band Adaptive Differential Pulse Code Modulation (SB-ADPCM), or HD voice for short.

Of course, the phones at both ends of a conversation need to apply the same codec or algorithm. An important function of Session Initiation Protocol (SIP) and Session Description Protocol (SDP) is to let the phones decide how they will communicate. More on them below.

In the future, if the needed audio bandwidth increases further, there may be reasonable solutions. The sub-band approach to encoding can draw on codecs that require lower bit rates so the number of sub-bands can increase as required. Faster LANs are migrating from 100 million bits per second (Mbit/s) Ethernet to 1 gigabit per second (1000 Mbit/s), so a call may find it has much more than a DS-0 available. Those LAN speeds are becoming easier to find in the WAN as well, with Carrier Ethernet and direct optical fiber access to the premises boosting access speeds to 10 Gbit/s or more.

2.4 LOW LATENCY

One of the great attractions of the circuit-switched telephone was the experience: it is almost like being there. For most of the calls made for both business and personal reasons, the latency across the network is so low that it is not noticed—unless the connection includes a satellite hop.

Even long-distance calls have minor delay when entirely on land lines. The PSTN has so many routes and switches that it is not necessary to divert calls far from the most direct path. (US carriers have been known to play games with the rules for collecting local termination charges by routing calls through Canada, but that's supposed to be a rare exception.)

Terrestrial copper cables and optical fibers propagate signals (including the bits that make up DS-0s and packets) at roughly two-thirds c, the speed of light in a vacuum, or $0.6\,c$. Microwave is faster, about $0.9\,c$ (air is not a vacuum), but lacks the deployed capacity of fiber. Microwave has fallen out of favor for long distances but is still popular to connect cell towers to switching centers in the backbone network, saving a few milliseconds over T-1 or optical backhaul.

The PSTN's circuit switches add only a few hundred microseconds of delay each, so most latency on LD is propagation delay. Historically a transcontinental connection of 3000 miles exhibited a delay of about 30 ms. With PBXs and phones adding almost no additional delay, the total was well below the threshold (150 ms each way) where the callers notice. To avoid a problem, the goal for round trip delay should be under 250 ms.

When latency is minimized and stable, it is constant. Thus in PSTN circuits there can be practically no variation in latency, thus no significant jitter. All the T&S equipment synchronizes to a master clock that ensures all DS-0 bit streams run at the same rate. In the worst case, an adjustment in TDM timing, called a frame slip, may drop or duplicate 125 microseconds (μs) of audio. A phone user might not notice at all; modems should not drop a connection.

The PSTN has been shifting to VoIP for long distance. Some carriers use the Internet, but major carriers separate the VoIP transmission facilities from the Internet and engineer them to minimize delays from congestion. Because the capacity of the backbone is so high—reaching 100 Gbit/s at this writing—the queuing delay in a router is even less than the few hundred microseconds in a 5ESS circuit switch. LD calls on the US PSTN continue to satisfy in terms of latency and jitter.

Cell phones have a completely different architecture. The radio link between the cell tower and the mobile handset (or tablet, or whatever) operates under a tight bandwidth constraint. After paying huge sums in bidding for RF spectrum, the carriers cram as many calls as they can into the resource. This means compressing the voice signal, oversubscribing the air side capacity, and suppressing transmission during silence to free the spectrum. All that processing takes time, which is noticed on a conversation as a tendency for both parties to speak at the same time, not knowing that the other was also speaking. Voice collisions are the main problem with high-latency connections.

2.5 CALL SETUP DELAYS

How fast a call connects is a valuable performance metric. Delays between the end of dialing and ringing at the called phone are very brief on most PSTN connections.

Circuit switches in the CO have reduced two components of delay to the point where most callers don't recognize they exist:

1. Time to receive dial tone. For decades most users have expected almost instant dial tone.
2. Time to complete a call and hear ring-back tones after dialing the last digit (post-dial delay). Problems may arise with VoIP if undersized servers handle call signaling.

The first depends on the CO switch or PBX to detect a phone going off-hook (caller picks up the phone) and assign a dialed digits receiver to that line. For urban and suburban service this delay seldom exceeds one second. It is so reliable that most people don't listen or wait for dial tone but assume they can dial immediately—and they can. When human operators responded to off-hook phones it could take many seconds to hear "Number Please."

The second delay, call completion time, is down to near zero for a local call and under 3 seconds for long distance connections. The North American phone system routinely beats the ITU's global target (set in E.721) of 3 seconds for local and 5 seconds for toll calls (LD) between end of dialing and ringing.

Callers can thank the signaling system (SS7) for the speed. The originating switch forwards a call request to the terminating switch in a compact packet format on a separate high-speed data network. That terminating switch checks if the called line is free or busy and reports back before either switch starts to set up a voice path. If the called party is free, the switches signal each other to reserve trunks and switch capacity for a call. If busy, the reply generates a busy signal and there is no load placed on voice-path trunks.

SS7 improved on earlier signaling in several ways. Analog CO switches signaled each other with tones on voice paths. To have a signaling path, the network started to build a connection as the caller dialed. The caller wouldn't know a called line was busy until the full voice path had been set up, wasting resources.

Packet voice, VoIP, has similar delays, but introduced by different sources. An IP phone plays dial tone to a caller after the phone gets the attention of the call control server. If the phone has been inactive or is just plugged in, it may have to register or authenticate itself before receiving an acknowledgment. An active phone should have an open TCP connection to the control server and be able to play dial tone immediately.

Call completion requires the call control servers to find the called device and determine its state: busy, free, not accepting calls, and so forth. This search may require multiple lookups in various databases such as DNS, ENUM, and SIP registry servers (see Index). With smart caches, the servers may have the needed routing information to locate the called party. There is another step where the end points, caller and callee, report to each other their capabilities and preferences for codecs and other features. A selection of the values to use for the session may be arbitrated by the call control server or negotiated directly between the end points. A more detailed discussion is located in the SIP, Megaco, and H.323 sections of this book.

2.6 IMPAIRMENTS CONTROLLED: ECHO, SINGING, DISTORTION, NOISE

In the bad old days, analog transmission had a few things in common with a vacuum tube hi-fi. Turn up the gain too high and you would hear noise. On long-distance calls, multiple amplifications were needed to keep the voice level up to where the listener could hear it. Analog amplification distorted the signal and added noise. Each amplifier also increased the noise introduced in earlier stages. Many repeats made the caller hard to understand.

If a little of the output leaks back to the input, an amplifier will oscillate: motorboating at low frequency or squealing at a high frequency. The same happened when uncanceled echo on analog telephone transmission lines was present in the forward direction. Singing was a feedback condition just short of oscillation that distorted the sound. To avoid these artifacts, network technicians tuned the gain of each amplifier and sometimes introduced deliberate loss to reduce echo.

Almost all these problems came under control with the digital revolution. Rather than amplify a voice wave form, the digital system receives distorted but recognizable pulses representing the 1's and 0's of the digital stream. Instead of amplifying pulses, the digital system generates fresh pulses—the distortion is removed at each regenerator station.

Echo, however, is (almost) always with us. When a caller's analog electrical wave hits some electrical discontinuity, part of the signal energy reflects back toward the source. If the system carries that reflected portion back to the origination, the caller hears an echo of his voice.

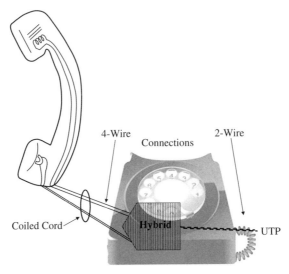

FIGURE 2.2 Telephone hybrid circuit joining the 2-wire local loop to separate pairs for the microphone and earphone in the handset.

Every POTS telephone has a built-in discontinuity, the "hybrid" circuit that connects the 2-wire local loop to the 4-wire interface on the handset: the earphone and the microphone have two wires each (Figure 2.2).

Echo from the speaker's own equipment is heard with almost no delay and is called "side tone." It assures the speaker that the phone is working. Most people don't recognize side tone until it is missing, which makes the phone line sound dead or out of order. Returning from the far end, echo is delayed by a time interval that depends on the equipment handling the call and the distance between participants.

IP phones, being computers, don't need hybrids. In most cases a packet-based phone need not create an echo, but that depends on the details of the phone's design. An exception is the speaker phone (hands free function). Echo cancellation (EC) allows the microphone and speaker to function at the same time, in full duplex mode, without creating a feedback loop that would generate a howl. The speaker phone knows what the received signal is, so it subtracts what looks like the received signal from the output of the microphone, before it is sent.

The quality of EC can distinguish one IP phone from another. Hardware phones designed for hands-free operation in a meeting room usually have very good EC. Normal desk phones may not have the same effectiveness in EC but can function well for one or two speakers in a quiet office.

Because analog phones will remain in service for a long time into the future, every phone system requires echo cancellation. Where large numbers of lines require echo cancellation, as in a central office, a dedicated product performs this function for an entire T-1 (24 channels), DS-3 (672 channels), or more connections. IP phones apply echo cancellation for one connection at a time in a software process.

3

FROM CIRCUITS
TO PACKETS

3.1 DATA AND SIGNALING PRECEDED VOICE

Data communications started with leased analog lines and modems, then later used digital services that ran at fixed bit rates. Mainframe computers supported thousands of terminals on shared lines that seldom exceeded 9600 bit/s—for 30 or more terminals per multidrop line. Packet switching brought greater flexibility to data transmission, and provided easier provisioning and troubleshooting.

3.1.1 X.25 Packet Data Service

The ITU started in the 1970s to define protocols and packet formats for accessing a packet data service. Under the umbrella Recommendation X.25, they defined the service interface between customer and network. How the network operated internally was left to each switch vendor. As was the custom before open standards became the norm, each make of X.25 switch added its own proprietary methods to register terminals, route connection requests, and create billing records. Frame relay standards followed a similar path, defining the user network interface (UNI) but not the network internals.

By the 1990s, X.25 service covered the world and was available on almost every computer made. It remained popular until after Internet access became widely

VoIP and Unified Communications: Internet Telephony and the
Future Voice Network, First Edition. William A. Flanagan
© 2012 John Wiley & Sons, Inc. Published 2012 by John Wiley & Sons, Inc.

available. The similarity between X.25 and SS7 allowed phone companies to offer a digital X.25 interface on ISDN lines, both in a DS-0 channel for user data and in the D channel (D stands for data, signaling not voice).

3.1.2 SS7: PSTN Signaling on Packets

Analog central office switches sent control information to each other over the same 4-wire analog trunks that carried voice. A 4-wire trunk uses a separate twisted pair for each direction, communicating in full-duplex mode. An analog switch put a high-pitched tone (2600 Hz) on a trunk to indicate it was idle. The adjacent switch did the same. A switch seized a trunk for a new call by going silent; the other switch went silent to acknowledge and to indicate it was ready to receive dialed digits. Even while residential phones had rotary dials, CO switches used tones for dialing because tones are quicker than pulses and easier to handle electronically.

When the phone phreaks (telephone hackers) realized how the PSTN worked, they built "Blue Boxes" to generate that 2600 Hz tone. They could trick the system into connecting to any phone number for the cost of a local call. It worked by placing a local call that involved more than one CO switch, say across town. The 2600 Hz tone played into the phone meant nothing to the first switch because it wasn't coming from a trunk. But the second switch was tricked into thinking the trunk from the first switch was idle. When the tone was turned off, the path through the first switch remained up but the second switch prepared to accept another call. The caller simply dialed another number using tones (DTMF). The digits reached the second switch—which assumed the first switch would generate a billing record. So the second switch routed the call to long distance without a thought to billing. The only billing record was created by the first switch, for a local call.

To prevent this form of toll fraud, the phone company separated control information from the voice paths. They migrated to a digital, packetized network called Signaling System 7 (the sixth and earlier systems were analog). SS7 (Figure 3.1) resembles X.25 packet switching, the standard on public data networks, but is different enough to avoid compatibility. Thus SS7 is isolated— phone phreaks can't hack into it. Thus was preserved civilization (and high rates for long distance calls, at least for a while).

SS7 had other advantages. It is faster, and doesn't involve voice trunks until the called end is known to be idle and able to receive the call. Switches use SS7 to access specialized servers that support features such as translating an 800 number to the directory number (DN, the unique identification of the called line) and providing location information about an E911 caller.

SS7 is very efficient, using compact messages where each bit has a meaning. Transmission lines among SS7 packet switches (signal transfer points, STPs) started at 56 kbit/s, and remained at T-1 into the late twentieth century. SS7 remains internal to telcos, with only modified access open to customers in the form of ISDN, Qsig, and a few other forms.

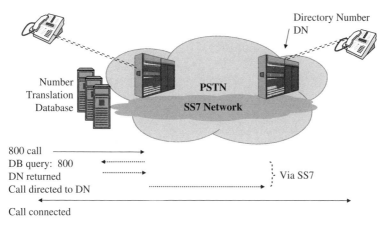

FIGURE 3.1 Signaling System 7, a packet switched data network that operates in parallel to voice paths among CO switches.

3.1.3 ISDN

Digital phone service in the form of integrated services digital network (ISDN) extended the 64 kbit/s channels (DS-0s) from the CO to the customer's PBX. Electronics multiplexed multiple voice channels onto one digital local loop. At the same time the telcos offered a digital-signaling interface to the PBX (the D channel) that made dialing and call handling faster. ISDN includes a time division multiplexed (TDM) signaling channel on the cable, the D channel, separate from the voice channels (B or bearer channels). Three forms were widely deployed:

- **BRI (basic rate interface):** two "bearer" channels for voice and a 16 kbit/s "data" channel for signaling, hence the shorthand designation of 2B+D. This format is used all over the world and remains the preferred access technology for residences and small business in Europe. Variants are offered, such as 1B+D for a single line phone and 0B+D for an always-on low-speed data connection used by credit card readers and similar equipment.
- **PRI (primary rate interface) on a T-1:** 23B+D appealed to enterprises for cost savings and remains popular with call centers because the D channel delivers caller ID.
- **PRI on E-1:** 30B+D fit in the digital line format used outside of North America. Synchronization and carrier management functions occupy one of the 32 time slots.

VoIP brings voice information and signaling back together on the same network, often on the same path for at least part of the way. That is, an IP phone has a single connection to the IP network, typically a LAN switch with uplinks to a core router. LAN switches carry IP packets without knowing the difference between them unless one type is prioritized ahead of the other by adding a field to the Ethernet header (IEEE 802.1p tagging).

Like phone phreaks, today's hackers can exploit access to the signal path to forge the calling party's identification, evade security measure by breaking out of a VPN, and eavesdrop on the conversations of others. Security remains an important issue.

3.2 PUTTING VOICE INTO PACKETS

This section will cover the basics of transporting voice (or video) in packets. The main difference between voice and video is the volume: voice has a constant rate of 40 to 156 kbit/s while higher quality video can use up to 2 megabits per second (Mbit/s), more for high-definition formats. Some video encoders will vary their output, depending on the rate of change in the image, so the amount of data per second can go down from the maximum.

Voice in packets has a long history. US Patents go back to at least the 1980s. Even earlier, the military used very low bit rate encoders on X.25 networks. These sounded awful but worked on the very slow connections then available and very few people knew how to tap it. One of the first public demonstrations of packet voice in a form practical for commercial enterprises occurred on a frame relay network at the Interop trade show in 1991 (by FastComm Communications).

As an historical note, the 5ESS™ electronic switching system was sold as a circuit switch. External trunks and lines were either TDM, transmitting voice in 64 kbit/s channels, or analog. Internally, however, the 5E packetized voice connections within the switch fabric.

The frame relay demonstration and the 5E operated with the same voice encoding developed for channel banks, pulse code modulation (PCM). It remains the standard for the PSTN. For VoIP within a LAN the encoding may be PCM or higher quality sound. On the WAN the option to compress the signal is often chosen to save expensive bandwidth.

To carry voice in a packet, the sending device saves up the constant rate bit stream for a short interval, then places those bits as the data payload in a packet (Figure 3.2). The receiver extracts the payloads and stitches them together again, recreating (for PCM) the 64 kbit/s stream and turning it into sound again.

VoIP and UC features such as real-time video function the same way. That is, rather than send digital information over a dedicated channel, the transport is in packet form.

Voice packets sent on packet transmission links reach the receiver at less precise intervals because of jitter or variation in latency across the packet network. A busy link will have to buffer some packets while sending others. A voice packet with high priority might have to wait for a large data packet to finish sending.

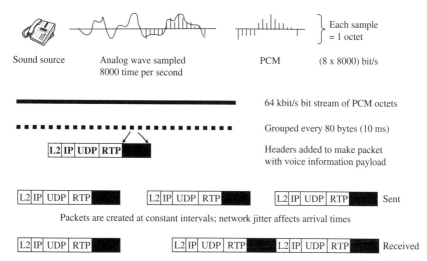

FIGURE 3.2 Continuous audio is digitized and the stream is divided into blocks for transport in packets.

To correct for jitter, the receiver stores incoming packets for a short time, in a jitter buffer, to ensure that the next block of content is available before starting to play back a packet. Buffering to remove jitter adds some latency, usually at least 10 ms. Adaptive buffering responds to the amount of jitter measured by examining the time stamps in packet headers. If jitter increases, the receiver may wait as much as 50 or 100 ms before starting to play back.

Silence suppression on the connection stops sending packets when the speaker is silent, allowing the receiver's jitter buffer to empty. Keep-alive packets carry time stamps, so the receiver can monitor jitter continuously, even during silence.

3.2.1 Voice Encoding

Much of the VoIP market relies on pulse code modulation (PCM, defined in ITU Recommendation G.711 and various Bell System publications). PCM is cheap to manufacture, simple, and passes not only voice but also modem signals, including facsimile transmission, with no additional requirements.

However, the addition of headers when carrying PCM in packets makes it a bandwidth hog, using as much as 156 kbit/s when encrypted. That's switch throughput, on both input and output ports. The recommended minimum bandwidth on a full-duplex access trunk (a clear channel T-1 carrying only IP on some data link protocol) is at least 80 kbit/s for PCM, 100 kbit/s if using Ethernet on the local loop. Vendors of VoIP equipment use two ways to encode voice that save bandwidth:

- Compress the voice payload (Figure 3.3).
- Suppress packets that carry no information other than background noise.

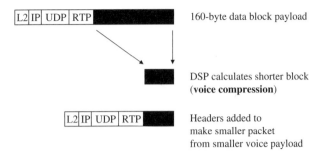

FIGURE 3.3 Voice compression shortens the payload but does not affect the headers.

An available feature of some decoders hides the effects of an occasional missing packet. The decoder stretches available information to cover for the lost packet. To the listener, this sounds better than collapsing the playback as if a piece had been cut from an audio tape. Any sound reduces the impact of the missing packet; the receiver bridges the gap with any of several forms of filler:

- Comfort noise resembling a generic background sound avoids the completely dead gap.
- A repeat of the last packet before the missing one. One missing packet typically represents only 20 ms so the listener may not notice.
- A sound synthesized by combining the content of the last packet before and the first to follow the lost packet.

Unfortunately for VoIP, packets tend to disappear in groups, rather than singly. The last method of replacing lost packets works better for lost bursts.

3.2.2 Dicing and Splicing Voice Streams

Figure 3.2 shows how voice input becomes data packets. First the sending end encodes the sound or video in digital form. Typically the first step for voice is the traditional standard of pulse code modulation (PCM); that is, a 16-bit analog-to-digital conversion, possibly mapped to an 8-bit scale. Some vendors offer no other form of encoding, although there are many. Some encoding methods offer better sound quality and higher fidelity. Some sample the input much more frequently—more than 40,000 times per second for audio CD's, for example.

To reduce bandwidth requirements, voice compression will process PCM to approximate the same information in fewer bits. For example, ADPCM encodes the changes between samples, rather than the full samples, in half or a quarter of PCM's bit rate.

Just as in a channel bank, any PCM encoder produces 8,000 digital bytes (octets or 8 bits) per second. The amount of overhead on a packet, and the limitation on packets per second that a router or switch can handle, make it impracticable to send each byte it its own packet. Instead, the sending device accumulates bytes to make a reasonably sized packet. In many implementations, 10 or 20 ms of samples go into each packet. That's 80 or 160 bytes of data in the voice payload.

Encoded voice needs additional information to carry it over the network and allow the receiver to decode it properly. As seen in Figure 1.7, the protocol stack consists of multiple headers. You will often see references to the size of headers being 20 bytes, occasionally a few more. This number refers to the size of the IP header, only one. The other headers, as described later, will more than double the overhead. Headers can exceed 64 bytes total, the size of the payload, which can double the bandwidth required per voice connection.

At the receiving end, the sections of sound in the voice packet payloads are unloaded and spliced back into a continuous signal to reproduce the voice. For systems using PCM, this means reproducing the DS-0 bit stream and decoding it as if it had come from a TDM channel. Other encoding methods use different playback strategies.

3.2.3 The Latency Budget

If applying compression on voice connections that extend over long distances, the latency budget is an important consideration. Examine the options for compression algorithms in light of the delay each introduces (see Table 3.1). Because processing power constantly increases, and the possibility that a vendor will develop specialized hardware chips to accelerate compression, it is best to get actual delay figures for the equipment you are considering. The only rule of thumb is that lower bit rates (e.g., from G.273.1) generally take complex methods that require more CPU cycles. For a given processor, complexity means more delay.

For highest quality voice, the choice of compression may be important.

Latency is important for voice and for real-time video conferencing in several ways besides affecting the MOS. The longer the delay, the more important it is to cancel echoes from the far end. All-digital systems may not create echo, but a digital phone call to an analog phone certainly will. Echoes delayed more than about 50 ms must be canceled to avoid interfering with the speaker. Uncanceled echo delayed much more than 200 ms can make a voice channel unusable.

The ITU's international rule for latency says:

- **150 ms, maximum on the PSTN:** above this level, latency starts to interfere with conversation. Too long a time window between speakers allows both to start talking without realizing that the other is also speaking, leading to voice collisions.
- **200 ms, acceptable for a private WAN:** participants start utterances at the same time often enough to be bothersome.

TABLE 3.1   The latency budget—contributions to packet delay

Component	Location	Approximate Contribution, ms	Remarks
PCM encoding	Originating phone	1	8-bit samples
Packet accumulation		10 or 2030	G.711/PCM, G.729 CELP G.723.1 ACELP
Compression and look-ahead delay		≤ 17	Depends on hardware and algorithm
Queuing			Behind 1 data frame on:
		≤ 100	10BaseT LAN
		≤ 10	100BaseT LAN
		≤ 1	GigEnet
Clocking out 120-byte packet (serialization)	Every device's sending interface: phone, router, switch	1 0.1 0.01	10BaseT LAN 100BaseT LAN GigEnet
Propagation	Transmission links	10 250	per 1000 miles terrestrial per satellite hop
Accumulation at the receiver (deserialization)	Every device's receiving interface: phone, router, switch	1 0.1 0.01	10BaseT LAN 100BaseT LAN GigEnet
Decompression and packet ordering	Receiving phone	8 to 19	Decompression is usually simpler than compression
Jitter buffer		20 to 60	May vary dynamically
Playback		10 to 30	Same as accumulation time of original packet
Typical total latency	On premises Long-distance call	54 100 to 150	No compression 2000 miles with simple compression and average router count

Note: Propagation delay depends on the transmission medium. Satellite hops are mostly in space so the average speed approaches the speed of light, $c = 300$ Mm/s (186,282 miles/hour). Optical fiber with an index of refraction of 1.5 slows photons to about $0.66\,c$; twisted copper pairs transmits information at about $0.60\,c$.

- **Above 250 ms:** created by a single satellite connection. Any echo makes the channel very difficult if not impossible to use. Even if echo is canceled perfectly, high latency makes voice collisions inevitable.
- **Above 500 ms:** possible on a channel that takes two satellite hops. Speakers might need to take unusual steps to prevent voice collisions; for example, say "over" when finished to emulate the "press to talk" procedure of radios.

A third artifact of compression arises when processing introduces different latencies for the audio and video streams. The receiver then sees an image of a speaker whose lips and other motions are not synchronized with the sound. Lack of lip sync can be annoying or even disconcerting. Video conferencing equipment should compensate for differences in latency. If not, choose another codec for voice and/or video if you have options, to see if there is a combination pair with a better match in latency.

Some vendors choose to apply voice compression by default. Some allow the user to choose the type of encoding. Some insist that voice be compressed to save bandwidth. Some users turn off compression for better audio quality or to allow facsimile transmissions. If you have the time and inclination, experiment with different combinations for audio and video to see what works best for you. Consider bandwidth usage as well as sound and image quality.

Any voice compression has some noticeable impact on the sound quality, but usually not enough to prevent conversation. Facsimile machine modems, however, are much more sensitive to transmission quality and generally fail to operate well, if at all, over a compressed voice channel. Technically the audio quality of the compressed voice channel is not adequate to reproduce modem tones accurately. Dialup fax machines transmit page images as digital bit streams sent over a modem, which converts the bits to audible sounds in the frequency range of a voice channel.

For fax, there are specific protocols and methods (e.g., ITU recommendation T.38) that treat facsimile separately from voice. More on this appears later in this book.

4

PACKET
TRANSMISSION
AND SWITCHING

We saw in Figure 1.6 how a circuit switch reserves a path for a connection when a call is set up and releases that bandwidth and switch capacity when the call clears. Each call requires a specific amount of each resource. It is relatively easy to calculate the fixed number of calls that can exist on each part of the network at one time. That is, a transmission link contains a fixed number if DS-0 channels. The number of simultaneous backplane paths in a voice switch may limit the number of calls it can set up to far fewer than half the ports on the switch. When a switch reaches its capacity, the next call must get a busy signal.

Packet-switched networks operate very differently. Resources are shared across all users. Some users, or rather their packets, may get higher priority or be allowed to occupy more resources than others, but bandwidth, memory, and CPU power are shared.

Rather than dedicate resources like bandwidth to a path or connection, a packet switch creates routing and forwarding tables that control where to send packets after they arrive (Figure 4.1). Regardless of the packet format, each switch or router answers the same questions about every packet it receives:

- Where should it go? That is, on which outbound link.
- What processing does it need? A router or firewall might translate IP addresses; frame relay switches will change the DLCIs (link addresses).
- Should this packet go next or wait while another packet is sent first?

VoIP and Unified Communications: Internet Telephony and the
Future Voice Network, First Edition. William A. Flanagan
© 2012 John Wiley & Sons, Inc. Published 2012 by John Wiley & Sons, Inc.

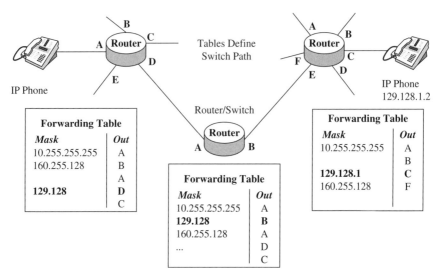

FIGURE 4.1 Routing and forwarding tables in packet switches determine where to send arriving packets.

Routers and switches share these functions. Differences arise because each type of packet handler makes its choices from analysis of different parts of the packet (Table 4.1). Usually the header(s) supply the information on which to make a routing decision. Some devices, such as load balancers, may look deeper, beyond the headers and into the payload, to examine values in a cookie or the header of a protocol such as HTTP.

Information of the type described in "Basis for Packet Handling" populates the forwarding table in each device. An L3 or IP router operates one or more routing protocols to determine which IP addresses are reachable on each link it connects to. MPLS switches examine only the LSP tags, so it is those values that occupy the forwarding table, as entered by RSVP or Label Distribution Protocol (LDP).

Taken together, the forwarding tables in the devices through which a packet travels are what define a virtual circuit (VC) for the packet connection. VCs may be:

- **Permanent:** configured by the carrier (or switch/router owner); comparable to a static IP address. All possible addresses are in the forwarding tables at all times.

- **Switched (connection-oriented):** The host that sends a packet must first notify the network by requesting a VC to a specific destination. Examples are MPLS, X.25, ATM, and frame relay SVCs for which the network inserts entries in forwarding tables only on demand.

- **Connectionless:** The network accepts a packet at any time, addressed to any recipient, and works out how and where to deliver that packet on-the-fly. IP routers operate in this mode, using information from DNS, ARP, and routing protocols to fill entries in forwarding tables.

TABLE 4.1   How packets are switched

Device	Basis for Packet Handling	Path-Finding Mechanism	Options
Ethernet switch	MAC addresses (layer 2)	Spanning Tree Protocol, configuration	Headers for priority, privacy, encryption
Internet router	IP addresses (layer 3)	Routing protocols: IGP, IS-IS, BGP, OSPF, etc.	NAT, prioritization (DSCP in TOS bits), encryption, etc.
MPLS core router	MPLS tag (in tag subheader) for each label-switched path (LSP)	Label Distribution Protocol, configuration for traffic engineering, RSVP	Prioritization, latency, encryption, etc., by configuration per LSP
Frame relay switch	DLCI (data link connection identifier)	Configuration, proprietary NMS, on command from end device (switched virtual circuit)	Traffic shaping via guaranteed bandwidth or b/w cap per DLCI; path selection for latency control
X.25 Switch	X.25 address in header	Proprietary, per switch maker	Closed user groups
Signaling System 7	Packet address (of switch or feature server)	Largely by configuration and internal telco process	Internal to carriers

4.1   THE PHYSICAL LAYER: TRANSMISSION

ISO's layer 1 (L1) is whatever carries the 1's and 0's. It's called the physical layer because in most instances L1 is a copper cable or a glass fiber—something physical, abbreviated PHY. Wireless LANs are called Wi-Fi based on a double pun:

1. The radio link carries 1's and 0's, so it's a wireless physical layer or Wireless PHY.
2. There is an allusion to the familiar Hi-Fi of high-fidelity stereo systems.

L1 could also be a light beam through the air—just another wavelength of electromagnetic radiation. Common examples of the PHY layer are category 5 or 6 LAN cable, T-1 links, DSL, and microwave transceivers. You can also think of the PHY as the connector level: RJ-45 (LAN), DB-25 (serial data), ST (optical), and so forth. The capacity of the available PHYs will be part of a network design review when preparing to add voice to an IP network.

Seldom does layer 1 impact the operations of higher protocol layers except for propagation latency. Consequently the PHY layer is considered outside the

IN SHORT: The Endian Wars

No scalps taken yet, more like an old feud, mostly worked out:

- Figures 4.2 showing that frame structures of protocols are read octet by octet, from left to right, with the row below following the row above.
- Representations are 4 octets wide because most protocols require PDUs and fields to be a multiple of 4 octets; also 4 octets is a computer word length on 32-bit processors. A PDU is padded at the end where necessary.
- The most significant bit is written to the left of an octet or a multi-octet field, which is the opposite of the order bits are transmitted on a serial line.

Issues arise in two ways:

1. How to assign bit positions within an octet? Two methods:

```
Bit Positions              8 7 6 5 4 3 2 1    ITU
                           0 1 2 3 4 5 6 7    IETF
                           +-+-+-+-+-+-+-+-+
Bit values in an octet     |1 0 1 0 1 0 1 0|
                           +-+-+-+-+-+-+-+-+
```

   The first "1" at the left, bit position 8/0, is the most significant and thus represents decimal value 128. Position $6/2 => $ decimal 32, $4/4 => 8$, $2/6 => 2$, for a total of 170 decimal. Putting the MSB to the left makes this a Big-Endian format or "network byte order."

2. What is the order of bits transmitted on a serial connection? Separate issue.

scope of this book but deserves a comment on the relationship between what happens "on the wire" and how it is written in documents. It's not obvious at all, due to a long-standing argument about transmission order: which bit in an octet is sent first, most or least significant?

Without trying to recount the history, an example will demonstrate. The bit string 0001 1011 is a binary octet written as transmitted in L to R order, with the least significant bit first. However, it is written in ITU standards as D8. The "half-octet" hexadecimal characters are written with the most significant digit to the left, as in Arabic numerals.

Fortunately, LAN sniffers and most test equipment will "decode" bits from the wire to present displays of protocol behavior in human readable form.

Early 8-bit computer hardware needed to handle only one octet at a time. To send those bits on a serial interface, chip makers settled on putting the LSB first—Little Endian. That order stuck as CPU's grew to 16, 32, and 64 bits and generally went Big Endian internally. Ethernet (IEEE 802.3) specifies that within an octet the least significant bit is sent first, the same order in which ITU numbers the bit positions. So the sequence for transmission, in general, is shown in Figure 4.2.

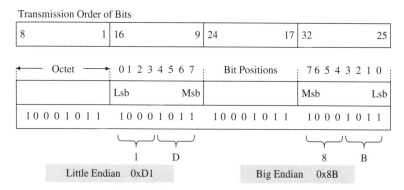

Transmission Order of Bits

| 8 | 1 | 16 | 9 | 24 | 17 | 32 | 25 |

FIGURE 4.2   Bit order and notation, Little vs. Big Endian.

This is true except for the frame check sequence, which is unique, and facsimile over IP. To preserve the order of bits transmitted between fax machines on the PSTN and the Internet, T.38 stipulates that the MSB goes first in the Internet Facsimile Protocol (IFP).

Fortunately, almost nobody needs to have a concern about bit order, but the circumstances indicate the level complexity that can exist when discussing communications.

4.2 DATA LINK PROTOCOLS

The data link, or layer 2, moves a packet from one host to an adjacent host, that is, over one transmission link. The most common L2 protocol in the enterprise is Ethernet (Figure 4.3). Standardized by the Institute for Electrical and Electronic Engineers (IEEE) as Standard 802.3, Enet has become the favorite L2 protocol because it works well, every equipment maker supports it, and it is both extensible and scalable. There have been so many additions that their numbering has gone through the alphabet (802.3a to 802.3z) and started over again (802.3aa . . .). Extensions important to VoIP include prioritization, authentication, privacy, and encryption. A given network may or may not use

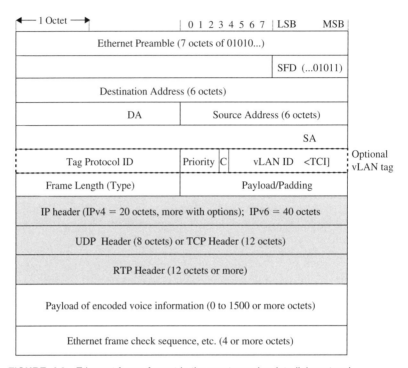

FIGURE 4.3   Ethernet frame format is the most popular data link protocol.

certain extensions and subheaders; to be certain of the format on a system, examine a datascope trace of actual traffic.

The Ethernet frame format contains source and destination addresses of the network interfaces on the hosts housing the sender and the targeted recipient. These are the Media Access Control (MAC) addresses burned into the Ethernet NICs. Each host builds a table of the MACs on its LAN segment using the Address Resolution Protocol. ARP allows a host to broadcast a request for a MAC address that matches a specific IP address. The host, switch, or router that can accept packets for that IP address will reply with the MAC address of its own NIC. Modern network designs based on switches reduce this list of MAC addresses to one per workstation, the port on the LAN switch serving this host. A hub in the network requires a host to learn all the MAC addresses on that hub.

Considering the size and complexity of the MAC header for Ethernet, in cases where each host has a dedicated switch port, this L2 protocol no longer performs most of its former functions. Originally Enet transmitted at 1 Mbit/s. Improvements in semiconductors allowed that speed to increase by a factor of 10 in each generation to 100 Gigabit/s—100,000 times faster.

Enet interfaces on routers and switches now support a single standard, type 1, which is a connectionless datagram service. Assured interoperability and increasing speed widened Enet's appeal to include almost any situation. In addition to LANs, for which it was designed, Enet now serves network access in the local loop because it can run directly over optical fiber and thus extend its reach far beyond the limitation of 100 meters on electrical cable.

That limitation came from the original L1 PHY, a coaxial cable. All stations on the LAN tapped in to the center conductor of a single cable. All stations shared the conductors so they heard all transmissions on the LAN segment. A station listened for a quiet state on the cable before transmitting. The preamble in the header was a warning to other stations that a packet was about to arrive.

Any other station that started a second preamble before hearing the first would hear that the cable was busy before the actual message started. If stations are much closer to each other—within 100 meters—the first will notice the second's preamble before it finishes its own preamble and will not start the header. When two stations start at the same time, their preambles collide (overlap and interfere with each other) allowing both stations to stop for short, randomly selected times before attempting to send again.

With full-duplex paths for faster Enet versions, and switches that isolate hosts on separate LAN segments, collision detection no longer operates this way. Today the distance limit can increase to reach a CO from a customer premise. Ethernet in the Last Mile Forum promotes "carrier Ethernet services" that use Enet as the L2 protocol on the access loop.

Legacy data applications (and some voice) use frame relay for the L2 protocol for carrier service outside of an enterprise LAN. That is, the local loop access link and the carrier's transmission network use a 4-byte FR header to create a permanent virtual circuit (PVC) from one customer site to another. Frame relay VCs as implemented by carriers are always permanent, which means they are provisioned by the carrier and the user has no control over their setup, routing, or tear down. Switched VCs (SVCs) are available in the frame relay switching equipment but have never been offered to customers. Analysts have speculated on reasons for not offering them, but carriers never gave a clear explanation.

4.3 IP, THE NETWORK PROTOCOL

Routers move packets based on the IP addresses, source and destination, in each packet. Every device needs an IP address. While this section was written, IANA announced that it had assigned the last available IP version 4 (IPv4) addresses to the regional Internet registrars (RIRs). They likely will assign all of them to ISPs by the time this book is printed. Hence the need to plan for IPv6.

Here's where the separation of protocol stacks into layers pays a big dividend. The change in IP will have very little effect on the data link and transport layers. Some applications may have to be modified to allow users to specify a choice of IP version and to enter the longer form of the new address.

IPv4 won't disappear for years, probably decades. But carrier networks may encourage users to migrate to IPv6 so that they don't have to run a "dual stack" of both versions in every device. There are defined ways to tunnel each version over a network of the other version. As used today, an IPv6 packet can be encapsulated in IPv4 to cross the Internet. As backbones adopt v6, the practice may reverse to encapsulating IPv4 in IPv6 packets.

Because of the potential that an enterprise will want to employ v6, at least to the Internet and carriers, every device should be selected with IPv6 capability, starting with those facing outward. NAT will allow v4 to persist inside the firewall, just as the 10.x.x.x and 192.168.x.x ranges of addresses do now. The next-generation NAT will translate between the two versions of IP.

Other external servers, such as firewalls; SIP proxies; and STUN, TURN, and web servers may run dual stacks for some time, as recommended in RFC 6157. But they will need to support v6 within their expected lifetimes.

To ensure the terminology is understood, this section covers both versions. The basic attributes of either version are that IP is:

- **Connectionless:** a sender can transmit an IP packet to any receiver at any time with no call setup; the network must cope with delivery, but only on a best-effort basis.
- **Not error protected:** IP will discard a packet that contains an error, or to relieve congestion.
- **Not sequenced:** IP doesn't guarantee packets will arrive in the order sent, but they usually do.

Figure 4.4 shows a version 4 Internet Protocol (IP) packet with its internal structure.

In practice, an IP packet never travels on its own. An IP packet, a layer 3 protocol packet, always has another, outer, header attached when crossing a

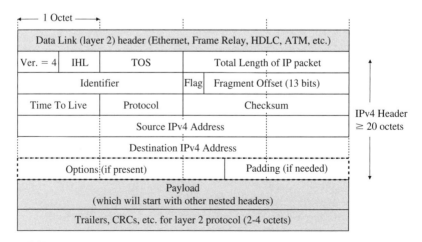

FIGURE 4.4 Packet layout for the Internet Protocol version 4 (IPv4).

transmission link. Every packet needs a header with an address associated with that data link or layer 2 in the ISO model for protocol stacks (see above). For example, IP packets on a LAN travel with a Media Access Control (MAC) header and MAC addresses assigned to Network Interface Cards (NICs) on Ethernet local area networks (LANs), as shown in Figure 4.3. On a wide area network (WAN) frame relay or the Point-to-Point Protocol (PPP) might manage a leased line link. On the local loop, other possibilities include Point-to-Point Protocol over Ethernet (PPPoE), Point-to-Point Protocol over ATM (PPPoA), and routed bridge encapsulation (RBE). The choice is usually determined by what the carrier will support for the service you choose. PPPoE (RFC 2516) often appears on Internet access links based on optical fiber or digital subscriber line (DSL) technologies.

On a frame relay network, the IP packet comes with a FR header that contains a FR address (data link connection identifier, DLCI) and control information. DLCIs identify virtual circuits (VCs) on that link and apply only between two adjacent switches. That is, a VC has one address that identifies it at both ends of the link (for a detailed description see reference 2).

Other, higher level, protocols represented by additional headers carried in the IP payload perform important functions on top of IP. For example, TCP not only discards errored packets, it also corrects those errors by retransmitting packets until they arrive correctly. The Real-time Transmission Protocol (RTP) carries sequence numbers that ensure that voice packets are decoded and played back in the correct order. These protocols are covered in Sections 4.4.1 and 4.5.1.

Fields in the IPv4 header identify the following parameters:

- **Ver:** version of the IP protocol, predominantly 4 until 2011 when those addresses reached exhaustion; being replaced by ver. 6.
- **IHL:** IP header length, the number of octets occupied by the header; needed so the receiver will know to process option fields when they are present.
- **TOS:** type of service, also known as the Class field; used to indicate expected handling to control priority, latency, or loss for this packet.
- **Total length:** number of octets in the entire IP packet, including nested headers, optional fields such as MPLS tags, and error check fields.
- **Time to Live:** every router that handles a packet decrements the TTL by 1; a packet is discarded if TTL reaches zero; prevents endlessly circulating packets.
- **Protocol:** indicates which type of header follows the IP header; UDP $= 17$.
- **Checksum:** a short number (16 bits) calculated by the sender from all the bits in the IP header; the receiver performs the same calculation and may discard the packet if the answer isn't the same (indicating an error in the transmission). The checksum does not include the data payload; the layer 4 or higher protocol checks for errors there.

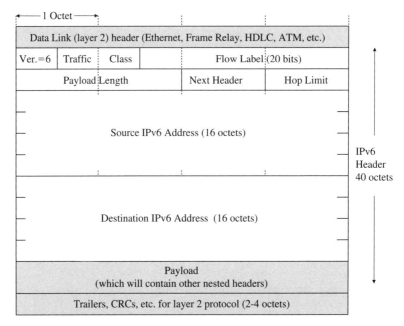

FIGURE 4.5   Packet layout for the Internet Protocol version 6.

- **Source address:** IP address of the sending host.
- **Destination address:** IP address of the recipient host. Either or both addresses may be changed by a router or firewall with Network Address Translation. This creates some problems for VoIP calling as discussed below.
- **Options:** control functions originally defined to help routers handle packets.

TOS and Options fields have evolved to indicate, for example, that a packet should move through the network with minimum latency because it contains encoded voice information.

To keep the Internet growing, IP version 6 offers addresses that are four times as long, in bytes. The IPv6 packet layout is shown in Figure 4.5.

Half the address octets are reserved for the host address, the individual machines on a LAN. In practice, this host address is based on the MAC address of the Network Interface Card (NIC). The manufacturer assigns the MAC address, which itself is intended to be globally unique, to each hardware port.

Including the MAC address means that a version 6 IP address may be tied to an individual device, not just to a network gateway as has been common to date.* To preserve privacy, there are ways for a host to pick a random number in place of the permanent MAC number.

The remaining 8 octets form the network address; 64 bits can express a number of network addresses that is the square of the maximum number of

---

* IP version 5 was never implemented.

IPv4 addresses. This is an astronomically large number that, with host addresses added on, should provide addresses for the indefinite future.

Simplification of the IP header in v6 kept the increase in header size to at most 20 octets for a minimal v4 header. If an IPv4 header carries options the difference in lengths is smaller.

Several fields were eliminated. "Next Header" is a name change for the earlier "Protocol" field but still identifies the first protocol header in the IP payload. Traffic Class indicates priority, formerly carried in the TOS bits. The new Flow Label can identify related packets in a stream, such as a video feed or a sequence of packets related to a transaction.

With the last IPv4 addresses in the United States allocated to service providers, the move to IPv6 is expected to accelerate. All VoIP equipment should be able to operate on an IPv6 network. IPv4 probably will be supported and necessary in many private networks for several years. Network Address Translation (NAT) will ease the migration to ver. 6 but also create problems for VoIP, as explained below.

4.4 LAYER 4 TRANSPORT PROTOCOLS

This layer in the protocol stack checks for errors in the data payload after it travels end-to-end and may exercise flow control to mitigate network congestion.

4.4.1 Transmission Control Protocol

TCP protects sessions against transmission errors and lost packets, but at a cost in memory usage, processor cycles, and potential added delay. It is "connection-oriented" because hosts that want to exchange packets must first open a session between them with an exchange of control packets (SYN and ACK). TCP is also "stream oriented" in dealing with octets rather than messages or transactions, per se.

To allow a correction, the sender holds packets until they are acknowledged as received OK. If no acknowledgment comes before a time-out period, or the receiver sends a retransmission request, the packet is sent again. These features make TCP a "reliable" protocol.

The TCP receiver uses the sequence number to identify missing data and to ensure the content from multiple packets is assembled in the correct order before that content reaches the software process handling that payload. The TCP header is shown in Figure 4.6. Note that RFC 793, from 1981, while still in effect at this writing, has been updated in RFC 1122.

TCP provides a multiplexing function to connect multiple host processes at a single IP address to other hosts with multiple processes behind one IP. The source and destination port numbers identify the process at each end. Some port numbers are "well known" from a permanent assignment by IANA to a

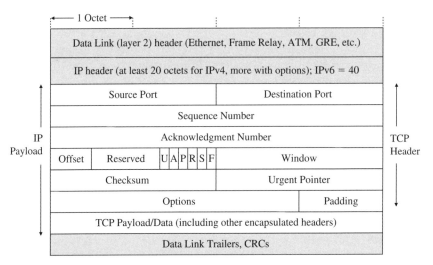

FIGURE 4.6 Transmission Control Protocol header layout. TCP maintains sessions so that it can retransmit packets when necessary to correct errors.

protocol; for example, port 80 is designated for HTTP. The six control bits coordinate the protocol between the two end points. Each is set to 1 to activate that function. The IP addresses and TCP port numbers at each end (four fields) define a session.

As a stream-oriented protocol, the sequence numbers apply to octets in a stream, not to packets. In effect every octet has a sequence number assigned to it by the sender. The initial number is random; then increments for each octet transmitted. ACKs use the sender's numbering to confirm safe receipt of octets to a certain point.

The basic functions of the TCP header fields are as follows:

- **Source port:** identifies the process at the host that sent this packet.
- **Destination port:** the process at the receiving host that will handle this packet.
- **Sequence number** (SN): the number associated with the first data octet in this payload.
- **Acknowledgment number** (AN): the sender's sequence number of the next octet the receiver expects, indicating that all octets with SNs before the AN were received OK.
- **Data offset:** the length of the TCP header (in multiples of 4 octets), indicating where the payload field starts.
- **Reserved**: must be all zeros.
- **U URG:** urgent pointer field is significant; if set = 1 the UPF indicates the SN of the first octet after the end of the urgent data (some of which may be sent in following packets).

- **A** ACK: acknowledgment field is significant; when set = 1, the AN field is acknowledging safe delivery of lower numbered octets.
- **P** PSH: push function; send immediately, don't accumulate a larger message.
- **R** RST: reset the connection.
- **S** SYN: synchronize sequence numbers or set up a connection (packet carries no data payload).
- **F** FIN: No more data from sender; closes connection (packet carries no data payload).
- **Window:** the number of octets that the sender of this packet is willing or able to accept from the recipient of this packet. Note that this 16-bit field will likely grow to 32-bits, so network equipment and terminals should be prepared to handle 32-bit window values or firmware updates in the future.
- **Checksum:** an error-finding calculation performed by the sender; if the receiver doesn't get the same value, the packet is dropped (and not acknowledged).
- **Urgent pointer:** if the U bit is set, indicates the end of the urgent data, some of which may remain to be sent in future packets.
- **Options:** extend the features or functions of TCP, such as enlarging the window field, defining a maximum packet size, or selectively acknowledging receipt of octets.
- **Padding:** may be necessary to place the start of the payload after a multiple of 4 octets.

The sequence number counts octets, but it lets the receiver recognize missing packets, either discarded because it contained an error or lost. In the basic procedure the receiver does not acknowledge receipt of subsequent octets until the missing ones are replaced. When basic TCP corrects a transmission error, it retransmits the missing octets and all subsequent packets as well—a waste of time and bandwidth. An option allows the receiver to selectively ACK (SACK) octets received after the lost packet, minimizing retransmissions. Options or header extensions in TCP follow the TLV pattern of type (2 octets), length in octets (2 octets), and value (some number of octets).

TCP is not recommended for voice, for several reasons:

- By the time a retransmission can finish, the packet is stale for real-time applications like voice and two-way video. The codec in the receiver can't wait while playing out real-time voice, so the correction packet arrives too late to be included and is discarded—more waste at both ends of the connection.
- For the sender to retransmit an errored packet, that packet must still be on hand. The sending TCP process discards packets only after the receiver

acknowledges accurate receipt. The additional memory and CPU processing needed to maintain TCP sessions imposes a significantly larger load on the equipment than does UDP.

- TCP slows down when it detects that packets are being lost, when it has to retransmit packets. This polite behavior was designed to respond to router congestion, and thereby reduce it. Interactive communications processes, particularly voice, operate at a constant bit rate. They need to send packets consistently at short intervals. If TCP were to slow the rate of packet launching, the connection would not be able to carry recognizable voice or smoothly moving images.

4.4.2   User Datagram Protocol

UDP provides the same multiplexing function as TCP, via source and destination port numbers, and end-to-end error detection. It is one of the earliest RFCs (768) still in use and an Internet standard.

The simplicity of UDP (Figure 4.7) means it requires less memory and processing power for the same volume of payload, so it introduces less latency.

UDP is a "fire and forget" process. It cannot correct transmission errors, but drops a bad packet. Hence UDP is an "unreliable, connectionless" protocol, also called as a datagram protocol (which IP is as well).

A system may also enable UDP to deliver "bad" payloads to the application layer, with a flag for the presence of an error. This behavior, configured outside of UDP, helps those video codecs that deal better with small errors, or partially damaged data, than with the loss of an entire packet.

The fields and their functions in the UDP protocol are:

- **Source port:** an identifier for the sending process of this packet and usually the port to be used for response messages.
- **Destination port:** the ID of the process to receive this packet.
- **Length:** the number of octets in the UDP header and payload.

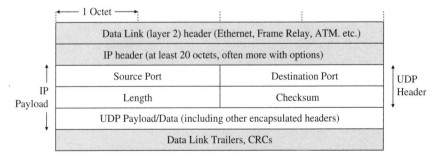

FIGURE 4.7 UDP header format and position in a packet. UDP offers "best-effort" service with no error correction.

- **Checksum:** a binary calculation that involves not only the UDP header and payload but also key fields in the IP header, including the IP addresses guards against delivery of a UDP protocol data unit (PDU, the UDP header and payload) to the wrong host.

4.4.3 Stream Control Transmission Protocol

SCTP is a multisession transport protocol developed to meet the needs associated with reliable transport of SS7 signaling messages over an IP datagram service. While designed for many forms of real-time transport, including voice and video, it has not been deployed much beyond a way to replace SS7 packet networks with IP networks. It is included here because of its potential applications in UC.

Like FTP, SCTP is reliable, but it offers more features that improve signal responsiveness:

- SCTP is message-oriented, so it can track the start and end of a block that gets segmented across multiple IP packets. This feature could be used by a receiver to confirm message boundaries, such as new master video frames.
- It allows multiple streams, multiplexing them on the connection, but enforces delivery sequencing only within a stream. Content in one stream might start later but arrive before content in another stream.
- Selective ACK (SACK) can acknowledge packets out-of-order, freeing up the sender's memory sooner.
- Data "chunks" from multiple streams can travel in the same IP packet, reducing header overhead.
- Multihoming allows a device on either end of a connection to be configured with two IP addresses for redundancy. Extensions allow adding IP addresses, dynamically, to support migrations.
- Inherent keep-alive functions detect lost links or failed remote hardware.
- SCTP has the ability to emulate a TDM circuit.
- Security is available via a protocol extension, Datagram Transport Layer Security (DTLS, RFC 6083).
- A "partial reliability" (PR) setting can be used to set the number of retransmissions.

SCTP appeared in 2000 as RFC 2960, obsoleted by RFC 4960 in 2007. In 2011 there were 22 IETF drafts active, and four RFCs published in 2010. SCTP is mandatory for SIGTRAN (Q.931 over IP), for SS7 transport in the PSTN, and for reliable server pooling (described in RFC 5351, informational only).

Cisco describes SCTP configurations to carry signaling (PBX format DPNSS and ISDN D-Channel messages) between a media gateway and the Voice Switch Service Module in a switch on the IP network. Alcatel-Lucent supports SCTP for backhauling SS7 and ISDN. AudioCodes uses the TDM emulation feature to make an IP network look like a leased-line BX.25 signaling network to the circuit switches (STPs) exchanging SS7 messages.

Despite considerable activity in refining and extending this protocol, there is no evidence that it is used as widely as RTP for VoIP and UC. One reason is lack of support in current NAT boxes; SCTP carries information that would require adjusting or rewriting deep in packets. At this time the response has reached the stage of an IETF draft document but not an RFP.

Error correction by selective retransmission and congestion avoidance are also part of SCTP. As explained elsewhere, retransmission isn't much help for voice and TCP works fairly well for SIP signaling. But graphical components such as white boarding or screen sharing could benefit from error correction. UDP is described in 3 pages, making it straight forward to implement; SCTPs RFC is 152 pages and there are other documents for extensions. Still its features are likely to be wanted by UC implementors when they know and understand SCTP better. Packing chunks from multiple streams into one packet is good for multiple voice channels between SIP endpoints. Setting PR to zero, emulating UDP to the extent of preventing retransmissions, would eliminate unusable repeats of voice packets.

The fixed and relatively low rate of data per voice channel prevents congestion of the kind seen when FTP ramps up the speed when delivering a large file. One voice codec presents a very limited load to the network. Other media could use all available bandwidth. To ensure that SCTP "plays fair" on the Internet, the protocol adopts the same congestion avoidance response as TCP, that is, rate dropping on packet loss and ramping back up gradually—still not desirable for voice.

The frame structure for SCTP starts with a common header, followed by "chunks" of data marked by subheaders (Figure 4.8). The port numbers and checksum work the same as in TCP. The verification tag is 0 in an INIT

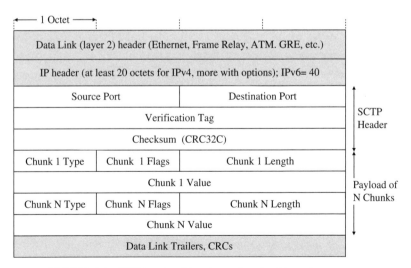

FIGURE 4.8   SCTP protocol header and payload format.

TABLE 4.2   SCTP message functions

ID Value	Chunk Type
0	DATA—payload data
1	INIT—initiation
2	INIT ACK—initiation acknowledgment
3	ACK—selective acknowledgment, maps received messages by transaction
4	HEARTBEAT—heartbeat request
5	HEARTBEAT ACK—heartbeat acknowledgment
6	ABORT—abort
7	SHUTDOWN—shutdown
8	SHUTDOWN-ACK—shutdown acknowledgment
9	ERROR—operation error
10	COOKIE ECHO—state cookie
11	COOKIE ACK—cookie acknowledgment
12	ECNE—reserved for explicit congestion notification echo
13	CWR—reserved for congestion window reduced
14	SHUTDOWN COMPLETE—shutdown complete
63, 127, 191, 155	Reserved for IETF-defined chunk extensions
Others	Available for additional extensions
	`0xC1 Address Configuration Change Chunk     (ASCONF)`
	`0x80 Address Configuration Acknowledgment (ASCONF-ACK)`

message; otherwise, the value in the response must match the value in the request. Verification tags can be the same for all messages in an association.

Defined chunk types (Table 4.2) leave many code points free for future extensions. Some have been reserved for projects started but not yet finished at the time of publication.

Chunk type numbers are assigned so that the two bits sent first in binary format (Table 4.3) tell the receiver what action to take if it doesn't recognize the type, which simplifies the parsing of this field.

TABLE 4.3   SCTP chunk classifications

00	Stop processing this SCTP packet and discard it immediately and silently.
01	Stop processing packet and discard it, but report an 'Unrecognized Chunk Type.'
10	Skip this chunk but continue processing other chunks in the packet; no report.
11	Skip this chunk and continue processing, but report the 'Unrecognized Chunk Type' cause in an ERROR chunk.

Using its own terms, SCTP creates an "association" between end points, each of which may have multiple NIC's and IP addresses linked over different paths. The association carriers up to 64K ($2^{**}16$ - 1) the port numbers for unidirectional "streams" or media channels.

A four-way handshake sets up an association, starting with a message containing an INIT chunk. To reduce the impact of flooding attacks, the recipient of INIT returns its information in a COOKIE chunk in an INIT-ACK message, then forgets about it. No resources are reserved and no state is saved. Only when the COOKIE chunk is returned by the caller does the second host send a COOKIE-ACK chunk and allow the association to add streams.

HEARTBEAT and HEARTBEAT-ACK message chunks keep a stream active when no user information is sent.

Closing an association normally takes a three-way handshake to provide confirmation both ways. The sequence starts with either end. Successive messages contain chunks of the types: Shutdown, Shutdown-Ack, Shutdown-Complete. An ABORT chunk ends an association immediately.

In almost all chunk types, the flags are set to 0 on transmit and ignored on receipt. In the ABORT and SHUTDOWN-COMPLETE chunks, bit 15 is set to 1 to indicate that the Verification Tag in the response message is not the responder's value but reflects the value in the received message.

To achieve high availability, an extension to SCTP for Dynamic Address Reconfiguration was defined in RFC 5061.

4.5   HIGHER LAYER PROCESSES

Voice packets are identical to data packets on a local area network (LAN) as far as the layer 4 headers. Differences lie mainly in the payloads of TCP or UDP, which contain additional headers for voice or video that are not used for data.

Each medium—voice, video, white board, instant messaging, and so forth—gets a separate communications session, each one identified by the UDP port in the header that precedes RTP in the packet.

The host applications must assign ports to streams, negotiate compatible formats, and synchronize received media where required, for example to maintain lip sync between a video of a talking head and the audio.

Separating media by session allows a receiver to choose among them, minimizing bandwidth consumption and unnecessary processing in the terminal.

4.5.1   RTP

In addition to voice information, each VoIP packet needs several headers to determine how the network handles that information. The first header added to the voice information is for the Real-time Transport Protocol (RTP), as shown in Figure 4.9.

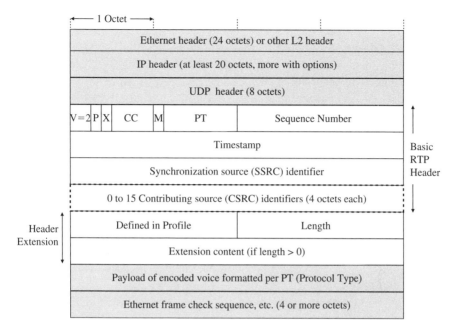

FIGURE 4.9 Real-time Transport Protocol for voice information.

The RTP header provides services that are designed for streaming media, such as voice and video, that require a constant-bit-rate connection. This header identifies the format of the payload, but that designation works in conjunctions with a separate profile of the application or system. The meanings of certain fields depend on how they are defined outside of the RTP specification (RFC 3550), for example, in the profile of RFC 3551, *RTP Profile for Audio and Video Conferences with Minimal Control*. Taken together, multiple RTP connections comprise a multimedia session.

RTP doesn't require any particular port number (for UDP or TCP) when sending and receiving. However, when communicating through a NAT firewall that is not aware of RTP at the application layer, RFC 4961 requires that a session send and receive on the same socket (IP + port number). This is called symmetric RTP, and it allows responses to pass the NAT because they are addressed to the source that opened the port in the firewall with an outbound packet.

This common behavior of a firewall, to open a port for return packets when a host inside sends a packet outside, may have limits applied. The firewall may not allow packets to enter the LAN if they are:

- Addressed to a host IP that opened a pinhole but not to the same port.
- Not from the destination IP and/or port address in the packet that opened the pinhole.

The RTP header fields have these functions:

- **V** $= 2$: the current version of RTP. Previous versions are obsolete.
- **P** $= 1$: indicates padding exists at the end of the payload; **P** $= 0$ means no padding. The last octet in the padding contains the number of pad octets to discard. This function should require no attention from the user.
- **X** $= 1$: indicates the presence of one variable-length header extension.
- **CC** (4 bits): the binary count, up to 15, of optional contributing source fields immediately after the SSRC field.
- **M:** a marker, set to 1 to convey any meaning defined in a profile for the system, for example the end of a video frame that occupies multiple packets. RFC 3551, the audio/video (A/V) profile, restricts its use to the first voice packet in a talkspurt following a period of suppressed packets due to VAD.
- **PT:** payload type indicates the payload format or how to interpret the payload at the receiver. Some PTs are defined in RFC 3551 but other profiles may apply.
- **SN:** the sequence number, starts with a random value and increments by 1 for each packet sent. The receiver can use the SN to detect lost packets and to ensure that received packets are played back it proper order. This SN increments differently from SNs in other headers such as IP and UDP.
- **Timestamp:** this clock allows the receiver to calculate jitter and synchronize the playback of different streams, for example, voice and video of a conference call. The value of the time can go out to multiple digits to the right of the decimal point, making it as precise as wanted.
- **SSRC:** the synchronization source identifier, a random number that should be unique within an RTP session. Every source of data (microphone, camera, file transfer process, etc.) has its own number to enable the receiver to keep related packets together when played back. A source that detects a collision with another having the same SSRC will cancel its number and generate a new oner. The SSRC is not intended to multiplex streams so a datascope reading of UDP ports should identify them clearly. A canonical name (CNAME) associated with a host may relate to many SSRCs.
- **CSRC:** contributing source, up to 15, represent sources of content that have been combined in some way, for example, other speakers on a conference call. A mixer will identify itself as the SSRC, and these sources in CSRC entries. A point-to-point connection will have no CSRC.

In addition to source and receiver hosts, RTP envisions specialized devices, mixers and translators, described in Section 4.5.2.

An optional header extension was inserted to allow vendors to experiment with proprietary functions. A receiver that does not implement this field simply ignores it. Use of the extension is discouraged in the RFC, particularly if another method exists that can produce the same result. Anticipating the potential for conflict among profiles, the standard requires that the first 16 bits of the header extension be unique to the function so the receiver will always be able to understand the meaning of the balance of the extension.

The current list of RTP parameters is maintained at `http://www.iana.org/assignments/rtp-parameters`. This list includes content from IETF drafts and new RFCs as well as the original RFC. There are more than 180 IETF documents about RTP.

Encryption of RTP, Secure RTP (sRTP), relies on a 'Crypto' attribute in SDP for the key exchange (RFC 4568). Both sides declare their supported cipher suites and attach the encryption key. Successful negotiation of encryption allows the call to proceed.

4.5.2 RTCP

A companion protocol to RTP, the Real Time Control Protocol (RTCP) provides monitoring and reporting for RTP connections. RTCP packets resemble RTP, but there are more alternate structures that depend on the type of RTCP packet. If examined with a datascope, all RTCP packets should be a multiple of 4 octets.

Like RTP, RTCP over UDP may send and receive on different port number—called an asymmetric connection (Figure 4.10). A symmetric RTCP has the same IP and port in a device for sending and receiving RTCP messages. The types are:

- **SR Sender Report**, which has three sections:
  1. Sender's SSRC and counts for report packets and the length of this packet.
  2. Sender's information about NTP and RTP time stamps plus counts of RTP packets and octets.
  3. Report on data received from one other SSRC, including lost packets, jitter, and delay.

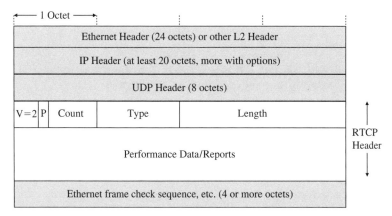

FIGURE 4.10 RTCP header has a structure similar to RTP but carries only reports.

- **RR Receiver Report**: same as SR but without item 1, saving 20 octets.
- **SDES**: Source DEScription, including CNAME the canonical name for a source.
- **BYE**: end of participation, should be the last message from an SSRC.
- **APP**: function defined for an application.

Each SR packet may report on data reception from up to 31 other sources on a multicast connection. For more than 31 sources, the host adds additional SR packets.

Often at least two RTCP packets are bundled into a single UDP data unit, to minimize packet processing overhead in routers. The first packet must be an RS or RR, even if no data have been sent or received. The second must show the CNAME, canonical name.

Each source should identify its CNAME to new participants as soon as possible, which is why RTCP was designed for each message to contain the CNAME in a SDES packet. The CNAME may associate with multiple sources of related media (voice, video, data).

A key function of RTCP is to sync the various clock and time stamps. Each media stream may have a different time stamp rate, and starts from a random point. At intervals a control message reports the various time stamp values at one point in "wall clock" time. All sites should be using a common time, such as NTP or GPS, to support synchronized media delivery and jitter calculations.

Counts of packets lost and octets sent or received allow a monitor of the connection to calculate the quality of the distribution network and estimate image quality as perceived by recipients.

The frequency that a participant host sends RTCP messages depends on the available bandwidth and the number of end points on a multicast connection. The goal is to use a small percentage of the network capacity for control messages, so the report interval should increase for more participants and/or lower link speeds. All stations must use the same value of the session band-width to calculate the report interval; the profile may set the bandwidth. If control messages exceed 5% of utilization, the configuration needs adjust-ment. To minimize bandwidth usage, mixers and translators should combine RTCP messages from multiple sources before forwarding.

The update RFC (5506) for RTP and RTCP allows a single packet per transmission under certain circumstances. Specifically, a host may send a feedback message alone, without constructing a full compound packet.

4.5.3 Multiplexing RTP and RTCP on One UDP Port

The original design of these two protocols separated them on adjacent UDP ports so that a carrier or service manager could monitor a private conference session or a multipoint video broadcast without receiving any of the content. In principle, muxing at the transport layer is good because it uses an existing facility (UDP) to simplify the processing at the next higher layer (RTP and RTCP).

Then came NAT. With address translation so common, the task of opening two pinholes per session became much more than twice as hard as one pinhole in a firewall. Some firewalls randomly assign ports to sessions. Such a device is incapable of assigning RTCP to the port one number above the port for the associated RTP session. That port might be assigned for another purpose.

Standards-track RFC 5761 describes how to assign RTCP packet types and RTP payload types so that applications can distinguish between the two protocols. The type fields in both headers occupy the same position in the UDP payload. By limiting the range of values used by each protocol, this field uniquely identifies RTP and RTCP.

When RTCP compound packets are used exclusively, only types 72 and 73 conflict. However, because another RFC (5506) allows single-element RTCP packets to save bandwidth, the recommendation is to follow the AV profile and avoid RTP type values, 64–95.

Before end points can multiplex RTP and RTCP, they must agree to do so in a signaling message. The attribute is indicated by a line in the SDP body of a SIP INVITE message: a=rtcp-mux. This attribute was registered with IANA.

If the SIP response contains the same attribute line, then muxing must be used; if not present in the response, the caller cannot multiplex. If a call forks, so that the INV is sent to multiple end points, the caller might receive multiple responses with and without this attribute. A caller who wants to multiplex must be able to mux for some endpoints and use separate ports for others.

Sending both protocols on one UDP port may interfere with header compression. An RTCP header isn't the expected change from the last RTP header so compression may not be applied to the packet. Smarter compressors can separate RTP from RTCP, the same way a receiver can, so that the two protocols can exist in separate contexts and experience full compression.

Muxing isn't recommended for the receivers to send RTCP messages.

4.5.4   RTP Mixers and Translators

The concept is fairly broad and can include transcoders that translate the video resolution or voice compression algorithm to accommodate participants on a conference call who connect over links with different bandwidths. Rather than restrict those with higher bandwidth access to a uniformly low quality, a mixer in the network can split A/V feeds into different streams at higher and lower bit rates. Participants on a smartphone might not be able to receive the full-rate video, and can't use it for display if they got it. Reducing the bit rate to them benefits all parties. The effect is similar to using layered encoding (Section 4.5.5).

Mixers may actually mix, as when combining video feeds from multiple sources into a composite screen that's fed back to all participants in a video conference. When combining inputs, the RTP header in output packets will carry the CSRC headers identifying the sources and the SSRC of the mixer. This information may be used by terminal applications to name portions of the display or attribute a speaker's voice to an individual.

A translator alters the payload in some way, but forwards packets without changing their SSRC fields.

Inputs and outputs at a mixer may be unicast or multicast. A translator can forward to a multicast address when receiving unicast packets, allowing one subscription to feed many end points.

4.5.5 Layered Encoding

Another way to offer flexibility in bandwidth usage is for the source to use a "layered" encoding in which the video image (the most likely medium) is separated into low-, medium-, and high-resolution information. Each component is transmitted in a separate RTP session, unicast or multicast. The receiver, knowing its own capacity to display an image, subscribes to those sessions which it can use. The end point consumes only as much bandwidth as it can process to the human user and does not waste bandwidth by discarding received information.

A cell phone with a small screen would connect only to the low-resolution stream, which would match the screens capability. A terminal in a corporate video conferencing center would take all available streams for maximum resolution. The same approach is possible for HiDef voice when it is encoded in sub-bands, each sub-band in a separate RTP session or multicast group. An audio end point subscribes to a PCM-quality stream and as many available refinement streams as it can use to improve sound quality.

4.5.6 Profiles for Audio and Video Conferences

In this context a profile defines the values for parameters described but not specified in the RFC for RTP. These include the codes for payload types and their mappings to encoding methods or algorithms. The profile completes the definition of an application so that instances from different vendors running on different operating systems and hardware may communicate successfully. The goal is to interoperate.

For audio/visual connections over RTP version 2, the starting point is often RFC 3551, *RTP Profile for Audio and Video Conferences with Minimal Control*, abbreviated as "RTP/AVP." Here we find the less frequent use of code numbers to represent parameter values for Payload Type, rather than text strings. You see something similar in the codes on video adapters that represent the PC's display resolution (e.g., VGA = 711).

After the code numbers were assigned, the introduction of Session Description Protocol (SDP) and other mechanisms provided a way to assign media codes dynamically, in an SDP block.

4.5.7 Security via Encryption

There are many ways to pick up copies of packets from a network at a point other than by a legitimate participant:

- In an IP multicast network, add a leaf station to the connection to receive packets; probably logged by a server and easy to trace.
- Instruct a router to copy packets to another address, legitimate for network diagnostics and a tool for CALEA, but also possibly for eavesdropping. Requires administrator access or a hacked router. With that access, an intruder should be able to delete or edit log files to delay detection.
- Install a bridge or tap on a link to copy arriving packets onto another port, often used for an intrusion prevention system (IPS) but also a potential opening for abuse. Requires physical access but may avoid detection by other than visual inspection.

To avoid these vulnerabilities and promote confidentiality, the best practice is to encrypt the payloads between end points. This configuration is relatively simple when both end points share a call control server and the calls remain "inside" the enterprise network. Both media and signaling may be encrypted as the call control server can participate in handling encrypted signal messages.

If the call goes off the LAN, encrypting the signaling will prevent an application-aware firewall from understanding when a call needs a port opened. The call may connect if RTP is symmetrical and the firewall behaves normally, that is, it opens a port for certain return packets when a host inside first sends a packet on that port. Responses may be accepted from any outside source replying to the host on that port, or the firewall may pass a returning packet only if its source socket is the same as the destination socket of the packet that opened the pinhole.

Signaling in the clear with encrypted media ensures privacy, but may reveal who's calling whom—potentially valuable information. The RFC allows for selective encryption of RTCP packets, concealing user identifications but leaving sender and receiver reports in the clear for monitoring by the carrier or a third party.

Any encryption algorithm will discourage the majority of casual snoopers. The best algorithm won't deter the spies with unlimited resources. The choice of how well to protect VoIP and other UC transmissions is another trade-off between the cost of protection and the value of what's protected.

At the time the RFC for RTP was published, 2003, encryption for the IP layer (IP Security) was also in development but not ready for prime time. Lacking a practical encryption at a lower layer, work proceded on Secure Real-time Transport Protocol (sRTP), which protects end to end (if desired), as does IPsec. Several RFPs on this topic are now obsoleted by newer ones, such as RFC 3711.

IN SHORT: Public Key Infrastructure (PKI)

In the public key infrastructure (PKI), each host gets a public key, published as widely as desired, and a private key to be kept confidential by the user. When a host encrypts a message or text with either key of the pair, only its corresponding partner key will decrypt it. The algorithm works for both encryption and digital signing.

For encrypting, the sender uses the recipient's public key. Only that recipient's private key will decrypt the message. As long as the recipient keeps his private key private, no one else can read the message.

Authentication works similarly, except that encryption applies to only a summary (a "hash") of the message using the sender's private key. If the sender's public key decrypts the signature, and a locally generated summary of the message matches the one in the signature, the recipient knows the message must have come from that sender and has not changed in transit. This process also provides nondeniability for legally binding signatures.

To protect the public key, part of it is a signature based on the private key of the certificate authority (CA) that issued the key pair. The CA's public key allows anyone to verify a sender's public key—that's what browsers do automatically when connecting a secured session to a server. It confirms that server's certificate against the certificates from established CAs built into the browser distribution.

Security provided by PKI relies on the difficulty of factoring the product of two very large prime numbers, the public and the private keys. As the names imply, the public key is published openly and is available to everyone. The private key is a secret held only by its owner. If the keys are large enough, the time to "crack" the key will be very long. The value of decoded information declines with time and when greatly delayed seldom has enough value to justify the cost of cracking.

For VoIP, the leading use of PKI at this writing is to authenticate signaling.

Certificates stolen from a CA, that is the values of both private and public keys, have been used to hack major corporate sites that had employed those keys. No security is perfect.

In summary, the RTP or RTCP payload is encrypted as a unit. However, each of the RTCP packets receives a random 32-bit number prepended to it before encryption. This "salting" of the data makes it harder to break the encryption, as does starting with random numbers for fields that increment and the SSRC and CSRC identifiers. This approach offers less than maximum security.

By encrypting the IP payload, the UDP and RTP headers are hidden, IPsec prevents the more efficient versiions of header compression. sRTP leaves the IP,

UDP, and RTP headers visible, to allow the greatest reduction from header compression.

ZRTP: Media Path Key Agreement for Unicast Secure RTP (RFC 6189, April 2011) is a protocol that allows end points to exchange one-time encryption keys over the media path, for example, using the socket established by SIP. That is, there is no impact on a firewall or NAT device. ZRTP uses the Diffie-Hellman scheme but without the need for a public key infrastructure (PKI), certificate authority servers, or permanent certificates. The method can use PKI and certificates if available. ZRTP also includes authentication methods to detect a man-in-the-middle attack. Authors on this RFC came from the ZPhone Project, Apple, and Avaya, indicating this approach to handling encryption may have major support and be significant in the future.

With more general techniques now more mature, current practices may encrypt at the IP level in the source, using a new IPsec header to route the payload.

Transport Layer Security (TLS) is a standardized way to apply PKI. It is the encryption technique most used by web browsers to encrypt communications to a web server, where it may be known by the name of its earlier version, Secure Socket Layer (SSL). TLS is secure and, in conjunction with other protocols for key management, allows hosts to create public and private keys on demand. TLS has proved reliable for commercial transactions and should be adequate for VoIP and UC. While most commonly applied to authenticating servers to clients, the same protocol will also authenticate properly equipped clients to servers.

Carriers offer encrypted transport within their networks, but that privacy feature may not include the access portion. US Federal agencies must encrypt data (and voice) on networks that leave government offices or networks, so agencies apply it on WANs outside their gateways or firewalls. Skype encrypts voice payloads between end points.

Signaling between VoIP call servers benefits from digital signing for authentication via the public key infrastructure (Figure 4.11).

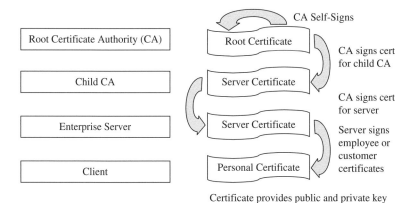

FIGURE 4.11 Chain of trust built by public key infrastructure from a Certificate Authority to a client application on a workstation.

Either form of PKI serves to authenticate users. If each person has a certificate, it can be used to encrypt or to authenticate packets and positively identifies the user so the data center can apply policy-based and role-based permissions. Encryption takes more processing power, so it often uses dedicated hardware to accelerate handling of large numbers of connections. Privacy often is less a concern, so authentication only, which requires less CPU power, can cost less to deploy.

4.6 SAVING BANDWIDTH

Clearly, the header overhead on packet voice is significant. There are several ways to reduce the bandwidth needed by each voice channel, including compression of headers and/or payload (Figure 4.12).

4.6.1 Voice Compression

Voices can contain frequencies upward of 10,000 Hz, but the region around 1,000 Hz carries the bulk of the information. Male voices have important content at even lower frequencies. Compared to the sampling interval of 1/8000 s, the sound wave at lower frequencies changes relatively slowly. As Figure 3.2 shows, a 1 kHz wave has about 4 samples per half cycle. That is, each measured point is usually close the previous point in value, creating redundancy. Analysis of the PCM encoding block of 80 or 160 bytes can recognize and remove the redundancy, transforming the encoding into a more compact format (Figure 4.12).

The payload size depends on the algorithm chosen. Table 4.4 summarizes some common choices. G.729 creates a 10 ms frame but defaults to two of these frames in a 20 ms packet. Compressing voice can reduce the 64 kbit/s stream

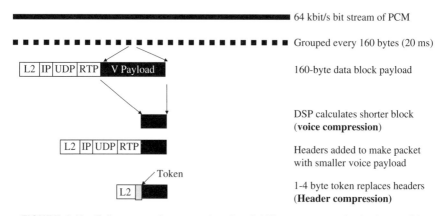

FIGURE 4.12 Full compression can reduce bandwidth per conversation by factor of 8.

TABLE 4.4   Voice payload sizes (in octets) for common codecs

Codec	Raw Bit Rate	10 ms Intervals	20 ms Intervals	30 ms Intervals
PCM	64 kbit/s	80 octets	160 octets	N.R.
G.726 (ADPCM)	16 kbit/s	20 octets	40 octets	N.R.
G.729 (CS-ACELP)	8 kbit/s	10 octets	20 octets	N.R.
G.723.1 (ACELP)	5.3 kbit/s	n/a	n/a	20 octets
G.722 (HD voice)	64 kbit/s	80 octets	160 octets	N.R.

Note: N.R. Not recommended: To control total latency, the ITU suggests accumulating voice for no more than 20 ms. G.723.1 requires an accumulation period of 30 ms, which increases the chance of exceeding the latency budget if used over the WAN.

to 32, 16, 8, or fewer kbit/s. However, the approximations necessarily become less accurate as the number of data bits declines, increasing the quantizing noise, which the listener hears as distortion. In the extreme, say at 1200 b/s, everyone sounds like Donald Duck.

Just how good, or bad, a voice connection sounds is measured in two ways:

- **Mean Opinion Score** (MOS): the average (arithmetic mean) of the scores that a team of trained listeners assigns to the sound reproduction quality of the voice channel.
- **Perceptual Speech Quality Measure** (PSQM): an algorithm (in ITU Recommendation P.861) that automates channel measurements without human judges.

Most often a vendor reports the MOS because the older metric is more familiar to most people. There are mappings between the two scores.

The MOS scale ranges from 5 (nearly perfect reproduction, like being face to face with the speaker) to 1 (unusable). Grade 5 was called "toll quality" because it was the target quality for transmission lines in the toll or long distance network among the end-office switches that served users' telephones. End-to-end scores include the local switches and the local loops; on a good day the MOS of a channel can be above 4. Most cell phones rate a 3 or 3.5.

VoIP may score low because an important factor in the calculation (or opinion) is the latency. VoIP packets have many opportunities for delay, including compression. Lower bit rates for the voice stream, via compression of the voice information, thus can harm sound quality in two ways:

- Reducing accuracy of reproduction through increased quantizing noise.
- Adding delay.

Computer processors, even dedicated digital signal processors (DSPs), need time to compress the voice payload. The receiver also takes a few milliseconds to decompress the payloads. Thus compression adds to the delay between when the speaker utters a sound and when that sound reaches the listener. The latency budget will be addressed in several ways, but for now we can see that processing time (5 to 20 ms) adds to the accumulation interval (10, 20, or 30 ms) to delay the packet. As might be expected, more complex algorithms that produce larger reductions in bandwidth tend to take longer to process. A summary of these and other delay elements appear in Table 3.1.

4.6.2 Header Compression

The inner headers for VoIP (RTP headers) function only at the end points, the phones. A connection could carry them with no significant change from packet to packet, over and over again, except for incrementing sequence numbers and time stamps. Replacing the longer string of header information with a much shorter token recognized by both ends eliminates redundant data (Figure 4.13).

The sender and receiver establish the call with full headers, then negotiate a token value as short as one octet to replace the header information. Earlier IP Header Compression (IPHC) (RFCs 2507 and 2508) used a 4-byte token. The sender inserts the token and marks the packet in the L2 header to indicate compressed headers follow. The receiver recognizes the header compression, replacing the token with the full header set before processing the packet. The receiver calculates the values of fields that increment, based on the rules for each protocol.

In some situations, the RObust Header Compression (ROHC, RFC 3095) Protocol relies on a protocol other than IP or UDP to identify packets whose headers are compressed. This implicit identification could be related to a range of MAC addresses, Frame relay addresses, physical ports, and so forth. For its efficient compression and robustness in the face of packet loss, ROHC earned a place on cellular radio links between handsets and base stations. Those links are all assumed to use it. Header compression taken to extremes becomes "link optimization." Specialized hardware devices compress headers, text, data, and images as well as spoofing protocols to reduce chattiness. When WAN acceleration appliances optimize a link, separate header compression is unnecessary. Compression of the voice payload at the phone may be worthwhile as that is a specialized task that the appliance vendors perform in a generalized way.

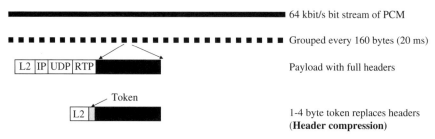

FIGURE 4.13 Header compression, which shortens packets but not payload, limiting the reduction in bandwidth.

TABLE 4.5 **Header sizes (octets) and effects of header compression**

Protocol	Full Headers	ROHC	VJHC
Ethernet[a]	28	28	28
IP	20	replaced	replaced
UDP	8	replaced	n/a
TCP	20	n/a	replaced
RTP	12	replaced	12
Replacement header	n/a	1 to 4[b]	4
Total octets IPv4	68	32	44
Header bandwidth[c]	27,200 bit/s	12,800 bit/s	17,600 bit/s
Added length of IPv6	20	0	0
Total octets IPv6	88	32	44
Header B/W IPv6[c]	35,200 bit/s	12,800 bit/s	17,600 bit/s

[a] Layer 2 protocol headers can be larger or smaller. For example, a VPN header may be added (to Ethernet) to separate voice traffic from other traffic. A frame relay header is as small as 4 octets.
[b] The token that replaces the headers may be as little as 1 octet; this table assumes a conservative 4 octets.
[c] Based on 20 ms interval, 50 pps; applies to each direction. Excludes voice payload which is additional and added at its nominal rate; e.g., PCM (64 kbit/s) with full headers = 99.2 kbit/s; G.729 with ROHC = 20,800 bit/s.

Van Jacobson header compression (VJHC) is an earlier form that replaces only the IP and TCP header combination with a 4-byte (or smaller) token (RFC 1144). Therefore it cannot apply to the preferred voice transport of UDP/IP. TCP should be used to ensure accurate delivery of signaling information. However, the volume of signaling traffic is much smaller than the volume of voice information, so it may not be worth the effort and complexity to configure VJHC. It functions only between adjacent routers, over one link, so each link must be configured individually. The resulting reduction in overall bandwidth utilization would be minor unless that link carried mostly signaling and not voice (Table 4.5).

4.6.3 Silence Suppression, VAD

The original application of PCM encoding in a channel bank used a continuous bit stream dedicated to each channel. The source (encoder) and destination (decoder) exchanged octets of voice information even when the circuit was idle, emulating the continuous current loop in analog phone trunks. This was a great simplification in channel bank design that cost nothing extra because the two copper pairs between channel banks, the T-1 line, have no other purpose. The idle condition wasn't always completely idle either because the CB's sometimes represented trunks that had to carry the idle-channel 2600 Hz signaling tone between CO switches. So each DS-0 was either in use for a call or handling signaling.

People don't speak continuously, however. In most cultures, one side listens while the other speaks (some families converse full-duplex, even on the phone). When a participant is listening, his "send side" of the connection carries no information and can be shut off. A technique to save bandwidth this way, time-assigned speech interpolation (TASI), has been around since the days of analog under-sea cables. If a cable carried 30 channels each way, electronic circuits might allow 50 conversations at a time, dynamically assigning a call to a channel only when a person was speaking. Digital versions took over later, with larger capacity and faster switching that allowed for greater oversubscription, up to 3:1.

Silence suppression in packet voice accomplishes the same oversubscription on a network shared by packets from many conversations. Some vendors call this feature voice activity detection (VAD). Others think of it as a phone claiming bandwidth on demand.

When a continuous DS-0 bit stream travels in packets, the "silent" packets occupy the same IP network bandwidth as active ones. Because the link is shared, these packets do impose an incremental cost on the system and can interfere with other conversations. That cost drops when the sender can recognize packets that contain no more than background noise and avoid sending them. In effect, the VoIP sender occupies bandwidth only when needed to send active speech.

Interrupting the flow of encoded voice information presents some difficulties for the receiver. It must decide if a lack of packets is just a lull due to VAD, missing packets, or a dropped call. The sender can help by sending short keep-alive packets that indicate the call is still up. Keep-alive packets may be sent at longer intervals than active voice packets, further cutting bandwidth usage. Receivers in commercial VoIP systems respond in at least two ways:

- Generate a generic "comfort noise" locally, perhaps side tone, so that the line does not sound "dead."
- Replay a snippet of an earlier period from the current call when the volume was at its lowest, presumably not speech. This approach sounds more realistic but may not be worth the complexity to implement.

To prevent user complaints, look for a VoIP system to apply one of these techniques.

RTP helps the receiver by marking each packet with a time stamp (wall clock time), which can preserve the intervals between utterances and the synchronization with the video component of a conference call. In addition the separate RTP packet serial number verifies that no packets are lost.

Laws of statistics apply to the possible saving in bandwidth. Just as with TASI, the more calls on a link, the better are the statistics and the more a design can allow overbooking that link. Another factor is how well the sender recognizes silence—the algorithm and the levels that trigger a decision determine how much the utilization drops.

The standard PCM encoding chip presents an electrical signal output when the encoded voice level is "silent." The sender may also analyze the PCM values over the interval accumulated for one voice packet, or multiple voice packets. As Figure 3.2 shows, the speaker's sound level varies during an utterance or talk spurt. There may be very brief silent periods and very brief sound periods.

To include short "islands" of sound after the main utterance, the sender may wait until the volume level over several packet intervals, of 10 or 20 ms each, averages to "silent."

To avoid excessive clipping of the start of an utterance, some sending systems can save enough raw sound input from "silent" periods to fill several packets. When a packet rises above the silent threshold, not only that packet but one or more packets from the intervals immediately preceding are sent in the proper order. Because the earlier packets are in hand, there is no additional latency to accumulate. The bit rate on the transmission line, always much higher than the bit rate of a voice channel, allows this group of packets to get off quickly, in a burst. The receiver easily resynchronizes the output.

By default, PCM receivers do not handle VAD packets. If they can, they will advertise the ability in the call set up information (by placing "payload type = 13" in the "m = audio" line of the SDP message, described below). G.729 voice compression in a receiver, by contrast, implies the ability to handle VAD unless the UA advertises "annexb=no" via SDP.

4.6.4 Sub-packet Multiplexing

To minimize latency, VoIP codecs generate a packet every 10 to 30 ms. The standard approach is to transmit them individually, each with complete headers (RTP, UDP, IP, and link protocol). In a device that handles dozens or hundreds of channels, such as a media gateway (MGW), most of the channels are encoded the same way, generating packets at the same intervals.

Methods exist that combine voice data from multiple channels into a single packet, saving many headers. Since the intervals are synchronized, multiplexing the voice channels adds little or no latency. The benefits are lower byte counts from headers (by a factor of 2 or more), despite additional subheaders inside the packet to identify channels. The big saving comes from fewer packets (a 98% reduction is possible).

The closest "standard" is in an implementation agreement from the Frame Relay Forum (now merged into other organizations): FRF.11—*Voice Over Frame Relay*. This IA describes an efficient format to combine channels with low header overhead. Once in a frame, the voice information can travel over an IP network in a layer 2 tunnel or directly as FR. The equipment that performs these functions generally hides the activity from SIP UAs, so they need no modification for this technique.

Proprietary methods are available. Dialogic offers ThroughPacket (TPKT), which multiplexes voice channels into an IP packet. Several vendors can encapsulate any HDLC frame into IP; HDLC includes the voice over frame relay format.

Both ends of the multiplexed connection of course must use the same scheme. In general, the greatest potential benefit arises on busy point-to-point paths. While the access link at headquarters may qualify, the remote office ends may not offer enough simultaneous calls to justify multiplexing voice sessions. Sometimes it pays to throw more bandwidth at a problem.

4.6.5 Protocol and Codec Selection

Some early implementations of VoIP (and one big-name software company currently) use TCP rather than UDP as the transport protocol (layer 4, the first header in the IP payload) for voice content. TCP is not appropriate for real-time voice transmission for multiple reasons. Always choose UDP for the voice path when a selection is possible. Many deployments default to UDP and should be left that way to minimize header overhead on links and processing on hosts.

Signaling connections among UAs and proxy servers are better served on TCP for several reasons. First, reliable delivery ensures that call attempts don't get stuck waiting for time-out intervals to expire. Second, the extensions to SIP and SDP expressed in the text-based format can boost the size of signaling packets above the size of the maximum transmission unit (MTU). When that happens, UDP has no ability to ensure that fragmentation and reassembly produce the original message—TCP does.

Modern voice codecs smooth over the short intervals left by an occasional missing packet. If the loss rate rises to approach 5%, however, the listener will hear it. When specifying equipment, consider how the receiver interpolates packet loss and the packet loss ratio for which it can compensate. Some systems tolerate packet losses better than others. If your network includes error-prone links (e.g., radio), the codec choice is important.

The User Datagram Protocol (UDP) is also known as the Unsequenced Data Protocol because the packets don't carry a time stamp or serial number (Figure 4.7). The sender simply pushes these packets onto the network and forgets about them. Routers read the IP header to direct the packet along its path. Any bit error will cause the layer 2 error check to fail and the router to discard the packet. There is no effort to correct errors, but neither is there time to retransmit packets that carry real-time voice or video.

The choices in local loops outside a building are moving to ever higher speeds. On a Gigabit/s link the added overhead of an Ethernet (layer 2) header of 24 octets may not be significant. If transporting many voice channels over a T-1 link (1.5 Mbit/s), it might make sense to save 20 octets per packet by compressing headers or configuring the link with frame relay connections that require no more than 4 octets per packet header. Modern routers are configurable with virtual routing and forwarding (VRF) instances that map a range of IP addresses to a frame relay address, a data link connection identifier (DLCI). This number, the same at both ends of the link, identifies the connection at both end points. It is significant only locally, so you can use the same DLCI at every remote location. Using the same DLCI simplifies preconfiguration of branch equipment before deployment and lets one spare device replace failed hardware at any remote location.

4.7 DIFFERENCES: CIRCUIT VERSUS PACKET SWITCHED

So far this chapter covered topics that most users recognize. In addition the following sections point out some peculiarities that differentiate packet technology from circuit switching.

4.7.1 Power to the Desktop Phone

Analog and digital phones in an enterprise draw power from the PBX. There a single backup battery keeps the phones working in a power outage. VoIP "switches" or call controllers provide signal and packet processing, not power.

In the place of the PBX, Ethernet LAN switches deliver power over Ethernet (PoE) (Table 4.6) not only to IP phones but also to Wi-Fi access points, video cameras, and other equipment. An existing data LAN might not be equipped for PoE, in which case phones may use line power via "wall wart" transformers. Soft phones running on notebook computer draw line power from the wall but also have some ability to continue on their batteries for a while during a power outage. The backup duration of a laptop battery can decline to near zero after a few years of constant use on local power. A machine might not reveal that its battery has lost its ability to hold a charge.

Availability is a larger issue with wall transformers. Battery backup is possible with a small uninterruptible power supply (UPS) at each cluster of phones, but phone backup is useless without backup for the switches, routers, and transmission equipment to maintain IP connectivity. Distributed battery backup can be a real chore to check and replace batteries every few years, and so on.

Where many phones exist in a small office area, transformers are bulky and messy. A PoE power insertion module in the wiring closet can provide the PoE function more neatly. A larger UPS for a PoE switch or power insertion module preserves both the LAN and phone services at a lower capital cost (for a given duration of support after an outage) and lower operating costs.

TABLE 4.6 Options to power IP phones

Source	Advantages	Disadvantages
Wall outlet	Cheap; generally available	No backup during outage; bulky, may crowd outlets where many phones exist together; adds another cable to the desktop
PoE from switch	Efficient, under smart control; allows central backup of many phones	Not available in older deployed switches; expensive initially; may require more power and cooling in wiring closet
PoE midspan inserter	Simple; delays replacing switches; allows central backup	Delays upgrade of switches; requires space in wiring closet; may not offer full control of power

4.7.2   Phone as Computer and Computer as Phone

Analog phones have as much intelligence as a light switch. Digital phones associated with digital PBXs contain electronics, probably a microprocessor, with behavior just above a clock radio.

A SIP phone without doubt is a computer. It maintains a protocol stack for communicating with other phones, runs applications, and displays variable images (context sensitive text or images, and going to video). Depending on the firmware or operating system, a virus or other malware can infect an IP phone.

PBXs are closed systems that run proprietary software and are seldom connected to the Internet. Creating and installing PBX viruses are difficult and, given the small amount of information stored there, seldom justify the effort to hack. Call servers for VoIP and UC not only run commonly deployed OSs, but at least some of them must connect to the Internet to send and receive calls. This exposure opens them to attacks, which must be prevented by means touched on elsewhere in this book but worthy of several books on their own.

4.7.3   Length of a Phone Line

Analog POTS lines run for miles in rural areas. A PBX port should be good for 3000 to 5000 feet. A design maximum on a local loop is 300 ohms of resistance rather than a specific length, which can be over 10,000 feet. The voltage drop across the resistance in the UTP copper conductors diminishes the sound level.

Digital phones draw power from a PBX over loops as long as 2 miles of copper cable. They use separate conductors in the cable and can double-up on wires to extend their reach.

Deploying VoIP to replace a PBX requires Ethernet connectivity at every phone. Slower versions of Enet limit the length of a "LAN segment" to 100 meters. A segment used to be the contiguous coax, now it refers to the length of a UTP cable between devices. The shorter reach of Ethernet can influence network topology (Figure 4.15).

Full-duplex Ethernet connections directly between devices (phone–switch, switch–switch) can exceed 100 m (Table 4.7) by using higher quality cable than specified for the LAN speed (e.g., using Category 6 on 100BaseT, not Cat 5 which is the minimum grade required). Extending copper cable increases the resistance in the power circuit, which may or may not affect phone operation because of lower voltage. Larger gage wire can help extend the reach of a cable.

Optical fiber can extend Ethernet for miles, but can't supply the d.c. current via Power over Ethernet (PoE), the most common way to power an IP phone. A fiber interface from router to switch allows placing the switch's PoE ports closer to the IP phones, but requires local line power and cooling at the switch. Analog phones need nothing more than a single copper pair.

Exercise caution when cables for PBX extensions go outside of a building. Standard port cards on switches and servers are not designed to survive surges from lightening strikes. Outside plant (cables not inside a building) need special overvoltage protection and lightning arresters (Figure 4.14). Even with such protection in place, a direct strike can burn out phones, circuit cards, or a chassis.

TABLE 4.7   Cable length limits for deploying telephones

Phone	Power Feed	Connection	Device	Distance[a]
Analog phone	Loop current	2-pair wire	Analog PBX	≥5000 ft[a]
Digital phone	Cable pair	4-pair wire	Digital PBX	<2 miles[a]
IP phone	Power over Ethernet	LAN cable	LAN switch	100 m
	PoE or line[b]	Multimode fiber	Optical line	2 km[c]
		Single-mode fiber	driver	20 km[c]

[a]For example: 22 AWG wire is needed for 5500 ft; reach is 3500 ft on more common 24 AWG loops; 2200 ft on is seldom used 26 AWG.
[b]Line driver may deliver PoE to phone.
[c]Cable between IP phone and line driver is limited to 100 m.

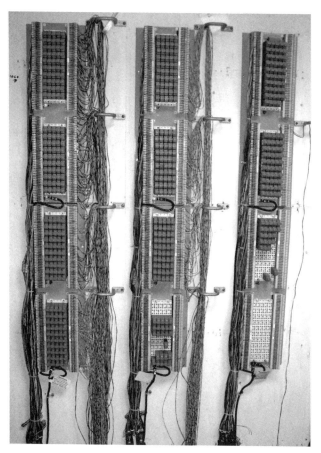

FIGURE 4.14 Over voltage protection for each pair required for outside plant, and the cables that extend beyond the building.

For a far remote device, there are media converters or line drivers that use fiber for the data. These can reach 2 km on multimode fiber and 20 km on single mode. One type of converter operates on a fiber cable and inserts power into the remote Ethernet port from a local line supply at that end. A second converter type draws line power from the "near" end, in the switch room, pushing PoE through copper conductors in a composite cable of fibers and wire pairs.

Media converters or line drivers can report loss of signal on either Enet or fiber, but are not highly manageable. This solution also consumes at least one and possibly two dark fibers. But fiber combined with a copper pair for power works well when a phone or camera at a distant and unpowered location, such as in a parking lot or by a gate.

Result: in campuses or large buildings the existing cabling, even if properly installed Category 5 or 6, may have a topology that won't work for VoIP (see Figure 4.15). An existing Cat 3 cable plant from a PBX will need to introduce active electronics (LAN switches) into distribution points that formerly held only passive cabling interconnections. Those locations will need a.c. line power, which shouldn't be difficult to install. But to keep the environment of the switch within operating specifications may require cooling, which can pose challenges.

In the past, telephone cable distribution panels could be placed almost anywhere wall space was available. A covered box offered enough protection. With a switch and air conditioning added, some box locations will need to add walls or dedicate a closet formerly shared with housekeeping services.

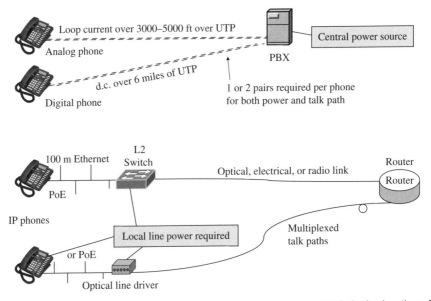

FIGURE 4.15 Length limits on cables affects network topology, particularly the location of power sources.

Construction and duct work can add to the cost of deploying VoIP. Older "pre-Cat" cables are unlikely to qualify for Fast Ethernet.

4.7.4 Scaling to Large Size

Small PBXs have a limit on the numbers of trunks and extensions supported that may be no more than a few dozen. Large PBXs typically allow the addition of cabinets to hold more physical ports on modules that may require additional shelves. At the high end, the capacity exceeds 100,000 phones.

VoIP systems, based on industry standard servers, expand by adding servers. A common ratio for a call control function is one server for no more than 5000 users. Features such as presence, access control, a directory, and management tools typically require separate servers at that phone count to provide sufficient capacity. A total of seven servers may be needed per 5000 users, more if the servers are to be fully redundant. The number of servers varies widely across vendors.

To improve availability, identical servers are clustered and fronted with a load balancer. To ensure adequate bandwidth and minimize latency, each cluster relies on a switched 1 Gigabit/s LAN. One vendor recommends a higher speed, either multiple 1 G or 10 G, among the many servers in a large installation. The required capacity of the LAN in the data center will be larger if the LAN carries transactions, storage, and backup traffic as well as voice and UC functions.

PSTN access from a circuit switched PBX favors ISDN primary interface (PRI) circuits because one D-channel signaling path can control multiples of the T-1 capacity links. Up to 28 PRIs require only a single DS-3 digital interface. A LEC routinely delivers large numbers of channels on an optical fiber formatted for SONET.

If a VoIP system reaches the PSTN through a gateway, the scaling up may involve a large capacity device or multiple devices under a gateway controller (GWC) or proxy server. Legacy PBX and CO switch vendors for a time had an advantage in experience with software control for large numbers of phones. The typical CO switch contained a dual-redundant processor that supported over 100,000 lines which terminated on other shelves. The architecture of the switch lent itself to breaking apart as it migrated to VoIP. For example, a redundant pair of just two powerful servers can control over 100,000 IP phone extensions. Additional pairs of servers are needed for other functions such as video conferencing and integration with email and instant messaging.

If provisioning VoIP for more than 50,000 phones, a server derived from a carrier switch may be the most cost effective because it minimizes the number of physical servers.

4.7.5 Software Ownership and Licenses

A PBX, once installed, typically runs indefinitely. The main ongoing expense is the labor to implement moves, adds, and changes. Periodic battery replacement isn't a huge issue. Damage from lightning should be preventable in most areas.

PBX software was closed and stable. Upgrades came years apart and were purchased separately, or might be included in a maintenance contract that also covered the hardware. One can live without a service contract and buy new software only when a new feature is compelling.

VoIP and UC fit a different model, the client-server computer system. Here changes to software not only add functionality but they also correct bugs and remove vulnerabilities. Upgrades are much more frequent than for a PBX "generic" (the software image) and probably are not optional—you need the new versions to stay safe and remain compatible with evolving standards and carrier services.

VoIP software vendors offer three main options:

- Software as a service or in the cloud. All hardware (except phones and terminals) and software are in the hands of a vendor who provides call control, messaging, and UC functions from its data center. The customer has only IP phones on its LAN. The cost is primarily the monthly recurring service charge, typically per phone, user, or DID. There is no explicit license fee.

- Licensed commercial software on hardware operated by the user organization. In some cases this is based on the number of server platforms, the number of CPU cores, or other data center metrics. The trend is away from one-time fees, the way games are sold, and toward an annual fee to use, forever, that may or may not include bug fixes and upgrades. In addition some vendors charge for a "client access license" for every authenticated user, for each application. "Pay as you grow" controls the cost per user, but removes the cap from total license fees.

- Open source software that incurs no license fee for the servers or clients on the user's servers. Enterprises may contract with value-added distributors who install, configure, and offer ongoing support for the system such as software upgrades. Vendors may sell customized editions of the general-public—license software or proprietary extensions to it.

IP phones, being computers, have a life expectancy on the order of a PC rather than a digital phone (and certainly not the life expectancy of an analog phone). Buying an IP phone with just enough features for each user role minimizes the capital expense. The least expensive IP phones at this writing seem to cost around $30. The "executive model" can exceed $500.

The $30 phone will likely avoid any software upgrades because it won't have the capacity, number of keys, or a display that the pipeline of new SIP features will require in the future. The cost to license a new software image and the labor to install it should make buying a new phone relatively attractive compared to buying a "future proof" model now for hundreds of dollars.

5

VoIP SIGNALING AND CALL PROCESSING

A fundamental assumption underlies the basic standards and recommendations for signaling in the IP environment: every device attaches directly to the open Internet and has a public, routable (i.e., easily reachable) address. Transactions to set up, modify, and end sessions—as defined in their primary documents—generally don't consider firewalls, network address translation, access controls, or other security measures that might exist at the boundary between the Internet and an enterprise LAN. Many of those security measures prevent VoIP signaling and conversation flows. Workarounds allow VoIP to work in practice on today's Internet, using IP version 4.

Examples abound of protocols that assume open connectivity: the basic RFCs for signaling protocols, the SIP Forum recommendation for SIP trunking, and most other descriptions of IP telephony. This section likewise will treat signaling first in isolation, to describe its functioning. A later section will describe how to communicate through the barriers set up by security measures.

There are three primary signal formats for VoIP:

- The oldest is H.323, a family of protocols created by the ITU with a goal of enabling the functions needed by carriers to provide not only a service similar to that of the PSTN but also one- and two-way video distribution.

VoIP and Unified Communications: Internet Telephony and the Future Voice Network, First Edition. William A. Flanagan
© 2012 John Wiley & Sons, Inc. Published 2012 by John Wiley & Sons, Inc.

- Session Initiation Protocol (SIP) is the latest version of signaling and has dominated development work since about 2000. All new products support it, and it attracts the most interest in the market. SIP came originally from the IETF and shows many characteristics of Internet formats such as HTML and XML. Eventually the ITU joined in the development and standardization of SIP, so it now is a joint responsibility of the IETF and ITU.
- Megaco and H.225 have narrower applicability: the management of media gateways (MGWs) that connect the PSTN or other circuit-switched voice networks to an IP or VoIP network. This protocol is used by both SIP and H.323 controllers. ITU has taken responsibility for Megaco. Some vendors predict SIP extensions will replace it.

None of these protocols is a complete system for VoIP or UC. They are components of a system that requires them and many additional protocols and formats to transport content such as voice, video, files, and collaboration activities.

This chapter will review these protocols individually. But first, a short review of what they have in common.

5.1 WHAT PACKET VOICE AND UC SYSTEMS SHARE

As successors to the PSTN, it's not surprising that all standards-based VoIP systems share many properties and features. First there are the basic elements required in a system (see also Table 5.1):

- Phones or terminals for video, white boarding, or other formats.
- Call control intelligence to route connection requests to the desired terminal.
- Databases that hold the locations of phones, registered users, permissions, policies, and authentication credentials.
- Means to connect packet-based networks to the PSTN or other circuit-switched voice system.
- Monitoring and management tools for media applications, including billing and accounting.
- A network infrastructure of switches with ancillary services and network management.

VoIP systems in general have two options for high-level topology.

1. The most efficient, and easiest, way to configure a VoIP service lets the network routing protocol (IGRP, IS-IS, etc.) select the paths for both signaling and voice "bearer" traffic. This will route the voice packets and some configuration messages (session description protocol, SDP, not contained in SIP requests or responses) directly between end points (Figure 5.1). Paths for signaling packets may differ from voice packets because signaling paths necessarily terminate on a call control server, which may be in a different location from the end points (phones or terminals). Of course, all paths may have to follow an access link between a branch site and the wide area network or Internet.

TABLE 5.1   Names of components in VoIP and UC

Component	SIP	Megaco	H.323
Customer' entity	User agent	Media gateway (MGW)	End point or terminal device
Route controller	Proxy server	MG controller	Gatekeeper
User database[a]	Registrar and location service	n/a	Gatekeeper's function

[a] Systems may draw information for authentication from backend databases operated by HR, for example.

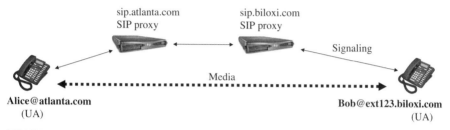

FIGURE 5.1   VoIP speech packets (bearer traffic) as typically following a direct path between end points such as IP phones. Signaling paths may terminate on a server. If end points can run a dual stack, one path may be IPv4 and the other IPv6.

Routing protocols establish one-way virtual paths. At times the two paths in a duplex connection may not fall on the same links and routers, which leads to asymmetrical traffic loads and difficulty in isolating faults.

2. Carriers may need to manage traffic for the purposes of billing, CALEA compliance to allow wiretaps, service management, or to ensure meeting network service level agreements (SLAs). For many reasons, carriers apply traffic engineering to bring voice traffic to specific points in their networks for monitoring or hand-off to other networks. For example, voice traffic may be routed through a session border controller (SBC) to an adjacent network.

Call control methods are in the class of request–response protocols. A device kicks off an action with a request to another device, which acknowledges with a response message. The procedures defined in the standards help to interpret these messages.

A few protocols associated with VoIP use messages that require no response, called indications. A keep-alive message to preserve a session is an example that fits into several places in the protocol stack.

These options are described in greater detail as part of the discussion below:

- SIP will receive the most exhaustive treatment as it dominates new products and appears destined to be the primary signaling protocol for Unified Communications as well as telephony.

- Megaco will be important as long as TDM telephony infrastructure provides any significant portion of the PSTN or a circuit-switched PBX.
- The explanation of H.323 is rather shorter. It is falling behind SIP in popularity, but H.323 exists in many locations and may persist for years.

5.2 SESSION INITIATION PROTOCOL (SIP)

SIP is a child of the Internet, specifically the Internet Engineering Task Force (IETF), which researches and drafts the standards for protocols and packet formats used on the Internet. It has a style that persists across time and functions. Of note here, the IETF likes control messages that contain text descriptions and messages that a human can read. For example, the address of a website is a domain name, even though the server is located by the IP address, a pure number.

Likewise HyperText Markup Language (HTML), the tool to create web pages, consists of text tags and variables which are sometimes cryptic but intelligible with practice. There is seldom any concern about the length of these control messages, or the size of packet headers. A criticism of this approach is that it assumes bandwidth is unlimited and free (see RFC5506, Section 1). This is not often the case except in a development lab where all the workstations are on a fast LAN. SIP messages often contain duplicated information in different fields, which can simplify processing of the packets at different points in the network. But the growth in message size from adding extensions means that a SIP signaling packet can exceed the size of the maximum transmission unit (MTU) on a link, forcing packet fragmentation.

Fortunately, enterprise LANs are getting faster, on average, each year and can exceed the capacity of the LAN in the lab that developed the protocol. Broadband access to WANs is getting faster too, so most users don't feel a bandwidth constraint for VoIP and UC.

Many sections of this book refer to a "SIP URI" or uniform resource identifier. In SIP, the URI contains enough information to create a session with a communications device or service, a "resource." SIPS does the same, but with the security of Transport Level Security (TLS), the successor to SSL. TLS is used for encrypted browser sessions that display a closed padlock in the status line. SIP and SIPS are called "schemes" in SIP.

The generic format for a URI is

```
[sip or sips]:user:password@[host name, FQDN, or
IP address]:port;uri-parameters?headers
```

If the host offers a service not associated with a user, the "user:password@" may be omitted. Certain characters are reserved, such as control characters and spaces. From RFC 2396, *Uniform Resource Identifiers (URI): Generic Syntax* we have the requirements for the URI:

> The set of characters actually reserved within any given URI component is defined by that component. In general, a character is reserved if the semantics

of the URI changes if the character is replaced with its escaped US-ASCII encoding [5]. Excluded US-ASCII characters (RFC 2396 [5]), such as space and control characters and characters used as URI delimiters, also MUST be escaped. URIs MUST NOT contain unescaped space and control characters.

To escape a character, replace it with a percentage sign (%) followed by one or two hexadecimal digits of its ASCII designation. An ampersands (&) is a separator. Some examples of valid URIs:

```
sip:alice@atlanta.com
sip:alice:secretword@atlanta.com;transport=tcp
sips:alice@atlanta.com?subject=project%20x&priority=urgent
sip:+1-212-555-1212:1234@gateway.com;user=phone
sips:1212@gateway.com
sip:alice@192.0.2.4
sip:atlanta.com;method=REGISTER?to=alice%40atlanta.com
sip:alice;day=tuesday@atlanta.com
```

As an application, SIP is independent of the network (IP version 4 or 6) and transport mechanisms (UDP, TCP; RTP, RTSP, SCTP, etc.). Likewise the sessions controlled by SIP may involve any media, formats, or protocols.

The RFCs, IEEE and EIA standards, ITU Recommendations, and various implementation agreements fully describe the procedures, conditions, exceptions, and options for VoIP and UC protocols. For further details, you may need the references listed in the Appendix.

5.2.1 SIP Architecture

From a high level, SIP consists of elements that communicate with each other using data link and transport protocols to enable procedures that set up, manage, and tear down communications sessions using a wide range of media. Some terms are defined specially for SIP, so this section will start with the key ideas.

SIP elements are the hardware and software in the VoIP or UC system:

- **User agent** (UA): IP phone, softphone, camera, video display, and so forth. Each end point incorporates the logically distinct components of UA client (UAC initiates requests) and UA server (UAS responds to requests). One UA software application usually acts as both client and server.
- **UA proxy:** a software server that acts as an intermediary, both client and server, to receive messages from and deliver messages to other proxies as well as to UA end points. Proxies provide discovery services when UAs need to locate a called party. An Authoritative Proxy has a public Internet presence and a DNS SRV resource record so that callers can find where they should send call requests for that domain.
- **Location service:** a database for a domain in which a proxy (e.g., the Authoritative Proxy) can find the specific host or UAS location of a called

party in that domain. The information comes from registrations that phones and other end points (UACs) make with a registrar server. The LS is a logical concept.

- **Registrar:** a server or process where an end point can post its location or address; a front end to a location service. A user may register an Address Of Record (a published or public SIP address) to be associated with multiple devices or UAs. More than one user may register at one end-point device.
- **Redirect server:** a proxy with access to information about location servers that can tell a UAC or proxy server where to send messages and perhaps offer suggestions for transport method (phone, fax, IRC), report a temporary or permanent move, and so forth.

How a network designer or software coder implements these functions is not specified in SIP. The different logical SIP servers may share a physical platform or software application. A common arrangement is two UAs in a device, each UA facing a different network. The back-to-back UA (B2BUA) can perform transcoding of media or signaling and provide security services. A session border controller (SBC) embodies this concept.

Small to medium organizations will likely deploy a single server (or two for higher availability). Large enterprises and service providers will create clusters of servers for higher capacity and availability. Specific servers can be optimized for different functions, installing more RAM or disk storage as appropriate for a call processor or a database server.

Transactions follow the model of request–response where one party (a SIP client) makes the request and the other party (a SIP server) responds. One application often acts as either client or server, depending on circumstances. The form and content of the messages depend on the type of transaction, called the "SIP method" (Table 5.2).

To help define SIP terminology, Figure 5.2 (based on RFC 3261) shows the sequence of the most significant messages and the main components in a call between two SIP phones. The INVITE reaches the called phone via their proxies; some message involve only the phones. This arrangement often exists in practice when the registrar and location service share a platform with the proxy.

Procedures for sending messages before a dialog exists differ from procedures within a dialog. Thus the INVITE message originates outside a dialog, but the OK response creates a dialog so that subsequent messages use the in-dialog procedures. A prime difference is that in a dialog, both UAs know the Tag value of the other and use it, with a Call-ID value, to identify the dialog.

The first message, an INVITE (Figure 5.3), is the request from Alice's UA client. The message goes to her local proxy, offloading from the phone the route discovery process to find Bob. Progress and status messages (Trying) flow back to the caller if the final "200 OK" message takes longer than 200 ms.

If Alice knows Bob's IP address, which could be saved from a previous call, the INV may be sent directly, without involving the proxies. Direct contact may be prohibited by the network administrator.

TABLE 5.2   SIP methods (message types)

Method	Abbrevations	Description	RFC
INVITE	INV	UA client starts a dialog and session. An INVITE sent within an existing dialog is called a re-INVITE.	3261
Cancel	CAN	Only for INVITE, and only after receipt of a provisional response.	
Acknowledgment	ACK	From calling UA client to called UA server after a final response.	
Register	REG	UA transmits location and other information to front end process of location service.	
Goodbye/quit	BYE	Always sent within a dialog (before that, use CANCEL) to close connection, end sessions and dialog.	
Options	OPT	To request and deliver specifics about the services supported by the other party in a dialog, or outside of a dialog without ringing.	
PRovisional ACK	PRACK	Used to ACK a provisional response.	3262
Update	UP	Updates session information before completion of INV-OK-ACK handshake without affecting dialog; can deliver identity securely or control "early media." A UAC should advertise UP in an "Allow" header.	3311
Refer	REFER	Transfer; USS asks UAC to issue a new INV to another UAS.	3515 5359
Message	MSG	Delivers Instant Messaging.	3428
Publish	PUB	Makes event state available, for example, presence.	3903
Info	INFO	Sends application data without changing state of session. Used for ISDN, DTMF, and Qsig signals, for example.	6086
Subscribe	SUB	Allows UA to ask for notification about some change.	3265 5359
Notify	NOTIFY	Informs subscribing UA of the change.	

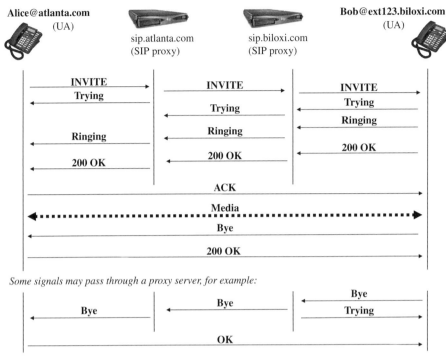

FIGURE 5.2 SIP call setup procedure, starting with an INVITE message.

```
INVITE sip:bob@biloxi.com SIP/2.0
Via: SIP/2.0/UDP pc33.atlanta.com;branch=z9hG4bK776asdhds
Max-Forwards: 70
To: Bob <sip:bob@biloxi.com>
From: Alice <sip:alice@atlanta.com>;tag=1928301774
Call-ID: a84b4c76e66710@pc33.atlanta.com
CSeq: 314159 INVITE
Contact: <sip:alice@pc33.atlanta.com>
Content-Type: application/sdp
Content-Length: 142
[                    CRLF for a blank line  ]
(Alice's SDP not shown)
```

FIGURE 5.3 SIP INVITE message to set up a call, showing mandatory headers.

The 200 OK from Bob's UA server is the response that confirms the connection to Alice's calling UAC. Alice's ACK confirms to Bob's UA that the connection exists, completing a three-way handshake that gives the UAC the information it needs to create a unique "dialog" with the UAS.

Note that the Bob's UAS adds a Tag parameter to the "To:" header. Further messages between these two end points (starting with the ACK) will be part of the dialog until the end points "hang up" with a BYE message.

```
SIP/2.0 200 OK
Via: SIP/2.0/UDP server10.biloxi.com
     ;branch=z9hG4bKnashds8;received=192.0.2.3
 Via: SIP/2.0/UDP bigbox3.site3.atlanta.com
     ;branch=z9hG4bK77ef4c2312983.1;received=192.0.2.2
 Via: SIP/2.0/UDP pc33.atlanta.com
     ;branch=z9hG4bK776asdhds ;received=192.0.2.1
 To: Bob <sip:bob@biloxi.com>;tag=a6c85cf
 From: Alice <sip:alice@atlanta.com>;tag=1928301774
 Call-ID: a84b4c76e66710@pc33.atlanta.com
 CSeq: 314159 INVITE
 Contact: <sip:bob@192.0.2.4>
 Content-Type: application/sdp
 Content-Length: 131
 [                          CRLF for a blank line    ]
             (Bob's SDP not shown)
```

FIGURE 5.4 OK response to an INVITE request.

The OK response that creates a dialog is in Figure 5.4. The UAC uses the Tag value in the To: header, added by the UAS, with its own Tag and the Call-ID to define a dialog.

Dialogs persist for as long as the parties continue to communicate. Additional messages about the call are part of the dialog, identified by the same three numbers: two Tags and the Call-ID. Either side may send requests for information, to change the parameters of a connection, and to add or delete media channels or sessions. The other side will respond to accept or reject the request.

The term "session" in SIP refers to a media connection between end users. It is logically separate from the dialog relationship that supports signaling transactions. One dialog controls any number of media sessions.

SIP needs additional supporting functions associated with any network. They are provided by DHCP and DNS servers, authentication controls, network access procedures, policy repositories, employee directories, and more. This book will refer to these functions but cannot cover them in detail.

The primary SIP RFC, 3261, divides the protocol structure into layers:

- Coding and syntax are the foundations of messages. These are highly technical and usually not visible to the user.
- Transport deals with sending and receiving messages. UAs and proxies select them automatically in almost all cases as they are standardized on the Internet.
- Transactions (requests and responses) are the application-level work of signaling performed by the transaction users. This is the area unique to SIP, the main subject of this section.
- Transaction user is any "core" (set of functions) within a user agent, registrar, or proxy server. This excludes stateless relays that don't send requests or terminate responses. This concept can help clarify processes but is not critical here.

TABLE 5.3   **Relationships of protocols in SIP system**

Session Format	Call Control	Call Status Discovery Methods	Encrypted Streams	A/V Streams	Conference Bridging
SDP	SIP SIPS SIP	RTCP (RFC3550) STUN	sRTP (RFC3711)	RTP (RFC3550)	SAP
TCP		UDP			
IP network layer packets (IPv4, IPv6, or both)					
Data link layer frames					
Physical layer connectivity					

The design of SIP minimizes the number of messages exchanged. For comparison, H.323 signaling involves many more messages, earning it the appellation of "chatty." To reduce the number of packets needed to set up a call, H.323 developed the "fast start" procedure that piggybacks negotiations for media and other parameters into the call setup messages. SIP takes this idea to completion: the Session Description Protocol (SDP) payload normally occupies the text field of the INVITE and the OK response messages. Negotiations complete without separate packets.

Because of their flexible text-based formats, SIP messages don't fit into clearly defined layouts of octets, like those used to describe the older protocol headers in IP, TCP, and UDP. Still it is possible to show which lower layer Internet protocols support the various SIP functions (Table 5.3).

Both TCP and UDP provide transport for SIP messages, depending on the context. For example, the request–response process for a phone to register with a proxy server, usually local to the phone, includes acknowledgments between phone and server that ensure delivery—the reliability of TCP is not needed. Thus UDP is an acceptable transport. INVITE messages to request a call may travel great distances, a situation where an end-to-end reliable protocol (TCP) offers benefits.

Like other protocols from the IETF, SIP is extensible. That is, the protocol allows optional fields, called headers (which are not the same as packet header in IP and TCP). These headers appear in the text block of requests and responses to label a specific parameter. A vendor can create a proprietary header, which can be named and filled with arbitrary text or values. To add a feature, for example, to a call request, it is not necessary to update the basic SIP standard for the INVITE message; a new header option can provide the needed information. A recipient that doesn't understand one of these options or extensions simply ignores it, ensuring that existing devices can interoperate with newer devices or those with future updates to firmware/software.

Each new telephone feature may have its own header and a separate RFC to define it. At one point there were over 50 groups working on drafts of technical standards for new SIP features. With so much change, it is comforting to know that most of the additions are likely to have only a small effect. To understand SIP, the broader functions of the standards will give you what you need to manage operational networks.

For a phone to be found by a caller, some infrastructure must exist (Figure 5.5) and be configured. Specifically:

- Each user has a SIP or SIPS (secure SIP) universal resource identifier (URI). These resemble an email address, but include the SIP label, for example, `sip:Dave@denver.com` and `sips:Ed@easton.gov`.
- The called phone must have a reachable proxy server whose IP address is known or findable, for example, through a DNS query.
- The called proxy must be able to find the called phone's current IP address.
- Registrars are contacted where the phone publishes its location (1 in Figure 5.5). Registrations may be permanent or for set times, after which the registrations expire.

When Dave dials the call to Ed, the User Agent client in his phone sends the INVITE message (3) to his local proxy. A phone may learn its proxy's IP from configuration or from DHCP.

Dave's proxy finds the address of the authoritative proxy for Ed's Address Of Record (AOR), `ed@easton.gov`, using DNS in the normal way. The query will ask specifically for the SIP server for the domain `sip.easton.gov`. These steps are not shown. With that address, Dave's proxy sends the Invite to Ed's SIP proxy (4).

A proxy that receives an INVITE, in general even the authoritative proxy, doesn't know the IP address of every phone in its domain. So when Ed's proxy, `sip.easton.gov`, receives the request for his specific end point, it

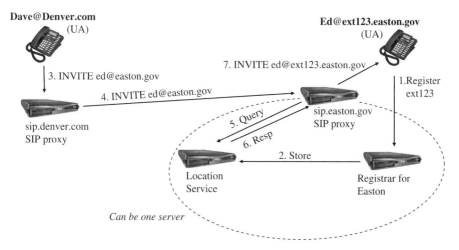

FIGURE 5.5 SIP REGISTRATION process and use.

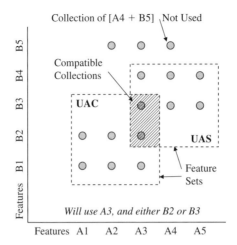

FIGURE 5.6   Compatibility between UAs defined as the overlap of feature sets.

may query the location service (5) and receive (6) the registered address, ed@ext123.easton.gov or an IP address.

Ed's proxy forwards the INVITE (7) to the registered phone. To make that address available, Ed's phone must have earlier registered (1) with the registrar server, which then stores that address (2) in whatever provides the location service. These logically distinct functions may exist as a single software application, on one machine. The location database may be part of the same application, but in large networks, or on a carrier service, the services may be on separate platforms optimized for each function.

If Ed answers, his phone responds with a SIP message, an OK response, that contains his direct IP address or URI and an SDP message about the codecs and other negotiable features it supports. The original INVITE contained similar information from Dave's phone. By following established selection rules (Figure 5.6) they both now know how and where to communicate with each other.

5.2.2 SIP Messages

SIP messages look like multiline paragraphs where each line is a header field that starts with an identifier, has some value, and may contain additional information or parameters added by the caller or an intermediary server. Table 5.4 is a closer look at the INVITE message from Figure 5.3.

UA clients need to match the requests they send with the responses received from servers. Various fields in the message headers provide the basis for a match, which fields depend on the message type. For example, the INV message contains a Tag in the From: header and a Call-ID. When the called UA server responds with an OK message, it adds a second tag in the To: header, chosen by the UAS. The two tags and the ID define a dialog, which can control many media session streams.

TABLE 5.4   SIP Request message format example for INVITE

Start Line	(INVITE sip:bob@biloxi.com SIP/2.0)
	method called address SIP version
Header Fields (others fields are optional or added elsewhere)	Via: SIP/2.0/UDP pc33.atlanta.com; branch=z9hG4bKnashds8
	Max-Forwards: 70
	To: Bob <sip:bob@biloxi.com>
	From: Alice <sip:alice@atlanta.com>; tag=1928301774
	Call-ID: a84b4c76e66710
	CSeq: 314159 INVITE
	Content-Type: application/sdp
	Content-Length: 142
	header type colon value semicolon parameters
Required Empty Line (marks end of headers)	[extra CRLF]
Message Body (optional text)	*(Alice's SDP goes here)*

Note: Each line ends in a carriage return/line feed. Example from RFC 3261. Content uses the UTF-8 character set.

The register method depends on the unique Reg-Id tag, Universal Resource Name (URN), and the registrant's Address Of Record to recognize an established "flow" between the proxy and the UA client. From that information, the proxy recognizes the return of a client when that client wants to change its address temporarily or update the contact information. The branch parameter in the Via: header may be used also.

SIP generally follows the standard format for Internet Messages (RFC 5322), which has the general layout of:

- Start line, indicating the method, protocol version, and so forth.
- Some number of header fields, each with a type and value, on separate lines
- Blank line (an extra CRLF is required)
- Optional message body; Content-Type and Content-Length headers define the format and size of this field

Clients generate Request messages that start with the method (INVITE, REGISTER, etc.) followed by the URI of the addressed entity (SIP: or SIPS:) and the protocol version. At this writing, the last must be exactly the string "SIP/2.0" for the current version. Header fields To:, From:, CSeq:, Call-ID:, Max-Forwards:, and Via:, are mandatory in all SIP requests, in addition to the start line.

A rule of request/response protocols like SIP is that one entity must finish a transaction it starts with a second entity before it can start another transaction of the same kind with the same peer.

Servers generate response messages whose start lines begin with "SIP/2.0" followed by a 3-digit code number and the text explanation (Table 5.5 is an example of a response 200 OK). Table 5.6 summarizes the types of response codes. There is another sequence of 3xx codes that are parameters for the Warning header in a response message; see the Appendix.

If a response indicates an error because part of the request is not supported, for example, 405 Method Not Allowed, then the response must list the supported methods in an Allow−Methods header. The same applies for Accept−Language, Accept−Encoding, and so forth. A list of headers and their functions appears in Table 5.8.

The response you want to see is 200: OK. If a UA client doesn't recognize a response code, the client treats it as if the code began with the same digit and was followed by 00. That is, if the client doesn't know 431, it act as if the server sent 400, Bad Request.

TABLE 5.5 Response message, 200 OK

Start Line	SIP/2.0 200 OK *version code explanation*

Header Fields
(others fields are
optional or
added elsewhere)
	Via: SIP/2.0/UDP server10.biloxi.com
	    ;branch=z9hG4bKnashds8;received=192.0.2.3
	Via: SIP/2.0/UDP bigbox3.site3.atlanta.com
	    ;branch=z9hG4bK77ef4c2312983.1
	    ;received=192.0.2.2
	Via: SIP/2.0/UDP pc33.atlanta.com
	    ;branch=z9hG4bK776asdhds
	    ;received=192.0.2.1
	To: Bob <sip:bob@biloxi.com>;tag=a6c85cf
	From: Alice <sip:alice@atlanta.com>;
	tag=1928301774
	Call-ID: a84b4c76e66710@pc33.atlanta.com
	CSeq: 314159 INVITE
	Contact: <sip:bob@192.0.2.4>
	Content-Type: application/sdp
	Content-Length: 131
	header type colon value semicolon parameters

Required Empty Line (marks end of headers)	[extra CRLF]
Message Body (optional text)	*(Bob's SDP goes here)*

TABLE 5.6   SIP status responses

3-Digit Code	Class	Explanation
1xx	Provisional	Request received, continuing to process the request
2xx	Success	Action was successfully received, understood, and accepted
3xx	Redirection	Further action needed to complete the request
4xx	Client error	Request contains bad syntax or can't be fulfilled at this server
5xx	Server error	Server failed to fulfill an apparently valid request
6xx	Global failure	Request can't be fulfilled anywhere

Source: RFC 3261.

SIP allows some header styles that do not conform strictly with RFC 5322, Internet Message Format, and the syntax and semantics of HTTP version 1.1 (RFC 2616). SIP is not an extension of HTTP but has its own transaction methods. In particular, they differ in handling white space when a header is folded over multiple lines.

UTF-8 encoding makes SIP messages easily readable in a datascope trace. As RFC 3261 states, headers will always be either:

- an opaque sequence of TEXT-UTF8 octets, or
- a combination of whitespace, tokens, separators, and quoted strings.

Text is insensitive to case unless quoted, as might be a "Value" in a header.

The INVITE method, one of several in SIP, starts with a line that contains the Request-URI, `sip:bob@biloxi.com` in this INV example. This field names the server that is expected to process the request.

Following the Start line are headers. Each header line starts with a field name followed by a colon and then the value. You may notice variable amounts of white space on either side of the colon—they don't matter.

- **Via:** identifies the location where Alice expects to receive responses and the transport method (e.g., UDP). Branch is a number set by the calling UA; it must begin with "z9hG4bK" to tell the server it is a globally unique value for each request (except a CANCEL). An optional rport= attribute either contains the source port in the public socket of this message or, if empty, requires the server to save that port value because it is needed for the destination address in response messages so the packet can pass a NAT firewall.
- **Max-Forwards:** limits the hop count until packet discard.
- **To:** a display name and the SIP or SIPS URI being called. May also be a TEL: URI for calling a public telephone number. The Tag in the To: header isn't present until added by the called party.

- **From:** display name and URI of the requester, possibly the caller's address of record. The Tag, a random string, is added by the phone to help identify the connection.
- **Call-ID:** globally unique for this call based on the caller's host name or IP and a cryptographically generated random number. The combination of the two tags (in To: and From:) and Call-ID defines a "dialog."
- **CSeq:** is a traditional sequence number, incrementing for each command message in a dialog. Also repeats the method type.
- **Contact:** only one direct route to the caller, possibly an IP address; not necessarily, but can be a URI; for future contacts.
- **Content-Type:** describes the message body, which is SDP in an INVITE. May be MIME-encoded, which requires a header for Content-Encoding.
- **Content-Length:** octets in the message body.

Each proxy that handles a request adds a Via: header with its own IP address and other data. When the called phone sends a response, the Via: headers act like source routing, allowing each proxy to forward the response immediately without a DNS lookup. When handling a response, each proxy removes its own Via header, so that when the message reaches the calling phone, it contains only that phone's Via, the state in which the request started.

Stateless proxies simply forward requests and responses without processing, other than its routing. Information about the message is not kept by the proxy. These servers add and remove Via: headers.

A home proxy or authoritative proxy receives INV requests for a public Address Of Record (AOR). That proxy looks up the registrations for that AOR, finds the URI that was in the Contact: header of the REG, and rewrites the message.

A proxy that forks an INVITE, that is, forwards to more than one destination, issues a 100 preliminary response or ensures reliable delivery. A forking proxy must maintain a state for the transaction including all dialogs and sessions. Both tags are needed to distinguish a two-party connection that results from a "forked" Invite sent to multiple UAs. Each UA that answers adds its own Tag, allowing the caller to sort out streams that otherwise share connection identifiers such as URIs and display names. A re-INVITE is always sent to a single URI, from a Contact header rather than the AOR, so it never forks.

Unless constrained by Record-Route (R-R) headers in the INVITE and response messages, the packets that carry voice, video, and other information follow normal routed paths for media. Adding R-R headers can direct all traffic through a specific IP address, for example, through the proxy servers if they are "watching" calls by maintaining a state machine for each call. Traffic could be routed through a call recorder, audio bridge, or a specific internetwork gateway if desired. If it remains in the path, a server must continue to proxy messages and forward packets until the BYE and OK messages that clear a call.

In general, the media packets are addressed to the Contact location after these end point addresses have been exchanged in the three-way call setup process: INVITE, OK, ACK.

A UA server may not be able to start a media session immediately. The UAS can keep a dialog alive and inform the caller of progress or status with provisional messages in the 1xx series. To prevent proxies from canceling a dialog for inactivity, a UAS sends a 1xx message every minute to avoid timing out while a call is on hold, for example.

After the series of notifications and status updates shown in Figure 5.2, the called phone or UA server responds with the 200 OK message similar to that in Table 5.5

Registration automates the creation of the location directory. A UAC sends a REGISTER message containing the standard headers plus its location (Table 5.7). Registration allows other UACs to locate the UAS for this user.

The headers for a Register message are as follows:

- **To:** address of record to be registered.
- **From:** AOR of person responsible for record; Tag created by the UAC.
- **Call-ID:** same value for all messages to this registrar to identify the UA.
- **CSeq:** incremented to ensure packer order.

TABLE 5.7   REGISTER method SIP message

| Start Line | REGISTER sip:biloxi.com SIP/2.0 |
	Method Register Domain SIP Version
Header Fields (others fields are optional or added elsewhere)	Via: SIP/2.0/TCP 192.0.2.2;branch=z9hG4bK-bad0ce-11-1036 Max-Forwards: 70 From: Bob <sip:bob@biloxi.com>;tag=d879h76 To: Bob <sip:bob@biloxi.com> Call-ID: 8921348ju72je840.204 CSeq: 1 REGISTER Supported: path, outbound Contact: <sip:line1@192.0.2.2;transport=tcp>; reg-id=1; +sip.instance="<urn:uuid:00000000-0000-1000-8000-000A95A0E128>" Expires: 3600 Content-Length: 0
Required Empty Line (marks end of headers)	[extra CRLF]
Message Body	[null in a REG message; Content-Length = 0]

- **Contact:** address(es) to bind to AOR; the REG-ID allows a connection to be restarted with the same conditions, for example after a link failure; the URN is globally unique and permanently associated with the device (e.g., a MAC address or random number).
- **Expires:** desired registration life in seconds, 32 bits.

The Contact: header contains the URIs or other addresses (IP, telephone, email, fax, etc.) where messages addressed to the Address Of Record should be sent. The UAC may add parameters to each URI to suggest when the binding should expire (the server has final say in some cases) and a preference (q=) for the order in which to use URIs. Asking for an "Expires = 0" should remove the binding.

If a UAC adds bindings to its AOR at a registrar, the response from that server will contain a Contact: header with all the addresses bound to the AOR. A REGISTER message without a contact header returns all the bindings in the response (200 OK).

To coordinate the wall clock time the UAC uses the Date: header in a request to get the time in the response.

To change a binding for an AOR, the UAC sends a REG message with the same URN and REG-ID: as an old binding, with a new URI in the Contact header where the AOR is reachable. A temporary move would carry an expiration time.

To maintain security, only SIPS addresses should be bound to a SIPS AOR. These will use TLS or another encryption scheme. If some other secure transport is in place, SIP URIs may be registered under a SIPS AOR.

5.2.3 SIP Header Fields and Behaviors

For full explanations of SIP headers in details that fall outside the scope of this book, start with RFC 3261, *SIP: Session Initiation Protocol*, and RFC 3840, *Indicating User Agent Capabilities in the Session Initiation Protocol*. They will direct you to other documents. See also the list of relevant RFCs in the Appendix. To study the logic behind the formats, see RFC 2533, *A Syntax for Describing Media Feature Sets*.

Table 5.8 is included for convenience in looking up headers. It summarizes SIP headers from several RFCs, the original SIP definition and later additions.

The terms "set" and "collection" have special meanings in Internet messages when using these message headers to announce capabilities or negotiate compatible parameters between end points.

- **Collection:** the operating conditions of a UA at any instant; the specific features in use then. A dot in Figure 5.6 represents a collection of two specific features that a UA can use together, one each from the A and B axes—for example, a codec and a bit rate.

- **Set:** all possible capabilities of a UA, all the values from the A and B axes supported by a device. The squares in the drawing indicate the capabilities of a specific UAS and a UAC: their feature sets. A UA includes its set when posting (REG) to a Registrar (location service) or in the SDP block of a negotiation message.

Two end points are compatible and can create sessions only if they share capabilities. That is, their feature sets must overlap. The cross-hatched area contains the compatibility region where the two UAs share features. From the collections within the shared set, the UAs choose one collection for a session.

If you want to stress your imagination, expand this two-dimensional figure to any number N dimensions with N representing the different kinds of features, such as audio codecs, video codecs, bit rates, image formats, and document types, each on its own dimension. A set then represents a volume in N-space, where overlapping volumes indicate compatible collections.

For less stress, think of the UAC and UAS each sending its shopping list of features and capabilities to the other. They do that in the Session Description Protocol (SDP) carried in the message bodies of INV and OK. Each UA compares the two lists. If any "items" appear on both lists, they can "buy" it for the call. If they have more than one to choose from, they look at each other's preferences for each item.

A ";q=" parameter (the "quality" of the choice, $0 \leq q \leq 1$) for each item on the list indicates preference: higher q value means more desirable. Using a common rule, they pick an item and use it to configure themselves for the call. If you were a gambler, or a statistician, you could consider the q value as related to the probability that an item is selected.

Table 5.8 and the following summary of headers that appear in SIP requests and responses, drawn from multiple sources, give some key points that might arise in examining a SIP system. Some of the most common headers have a "compact form" or abbreviation, shown in parentheses. They can shorten messages to avoid exceeding the MTU of the system.

Accept: which formats for content and media types can be handled. A header with an empty value means no formats are accepted.

Accept-Encoding: indicates an ability to handle forms of compression (gzip, tarz, etc.) and encryption. "Identity" means no encoding or the original format. In an INVITE, limits the encoding that can be applied by the UAS or proxy.

Accept-Language: the languages for code explanations, header texts, and so on. No such header indicates any language is acceptable. The "q = __" parameter can indicate preferred order among multiple languages.

Alert-Info: "alert" here means ringing a phone or otherwise calling attention of a called party to a new connection. Can be used to apply distinctive ringing by specifying a ringtone parameter.

TABLE 5.8   SIP message headers

Header	Where	Proxy	ACK	BYE	CAN	INV	OPT	REG	PRACK
Accept	R		-	o	-	o	m*	o	o
Accept	2xx		-	-	-	o	m*	o	-
Accept	415		-	c	-	c	c	c	c
Accept-Encoding	R		-	o	-	o	o	o	o
Accept-Encoding	2xx		-	-	-	o	m*	o	-
Accept-Encoding	415		-	c	-	c	c	c	c
Accept-Language	R		-	o	-	o	o	o	o
Accept-Language	2xx		-	-	-	o	m*	o	-
Accept-Language	415		-	c	-	c	c	c	c
Alert-Info	R	ar	-	-	-	o	-	-	-
Alert-Info	180	ar	-	-	-	o	-	-	-
Allow	R		-	o	-	o	o	o	o
Allow	2xx		-	o	-	m*	m*	o	o
Allow	r		-	o	-	o	o	o	o
Allow	405		-	m	-	m	m	m	m
Authentication-Info	2xx		-	o	-	o	o	o	o
Authorization	R		o	o	o	o	o	o	o
Call-ID	c	r	m	m	m	m	m	m	m
Call-Info		ar	-	-	-	o	o	o	-
Contact	R		o	-	-	m	o	o	-
Contact	1xx		-	-	-	o	-	-	-
Contact	2xx		-	-	-	m	o	o	-
Contact	3xx		d	-	o	-	o	o	o
Contact	485		-	o	-	o	o	o	o
Content-Disposition			o	o	-	o	o	o	o
Content-Encoding			o	o	-	o	o	o	o
Content-Language			o	o	-	o	o	o	o
Content-Length		ar	t	t	t	t	t	t	t
content-Type			*	*	-	*	*	*	*
CSeq	c	r	m	m	m	m	m	m	m
Date		a	o	o	o	o	o	o	o
Error-Info	300-699	a	-	o	o	o	o	o	o
Event		a							
Expires			-	-	-	o	-	o	-
From	c	r	m	m	m	m	m	m	m
In-Reply-To	R		-	-	-	o	-	-	-
Max-Forwards	R	amr	m	m	m	m	m	m	m
Min-Expires	423		-	-	-	-	-	m	-
MIME-Version			o	o	-	o	o	o	o
Organization		ar	-	-	-	o	o	o	-
Path	R	ar	-	-	-	-	-	o	?
Path	2xx		-	-	-	-	-	o	?
Priority	R	ar	-	-	-	o	-	-	-
Proxy-Authenticate	407	ar	-	m	-	m	m	m	m
Proxy-Authenticate	401	ar	-	o	o	o	o	o	o
Proxy-Authorization	R	dr	o	o	-	o	o	o	o
Proxy-Require	R	ar	-	o	-	o	o	o	o
RAck	R								

TABLE 5.8 (*Continued*)

Header	Where	Proxy	ACK	BYE	CAN	INV	OPT	REG	PRACK
Record-Route	R	ar	o	o	o	o	o	-	o
Record-Route	2xx,18x	mr	-	o	o	o	o	-	o
Reply-To			-	-	-	o	-	-	-
Require		ar	-	c	-	c	c	c	c
Retry-After	404, 413, 480, 486		-	o	o	o	o	o	o
	500, 503		-	o	o	o	o	o	o
	600, 603		-	o	o	o	o	o	o
Route	R	adr	c	c	c	c	c	c	c
Server	r		-	o	o	o	o	o	o
Subject	R		-	-	-	o	-	-	-
Supported	R		-	o	o	m*	o	o	o
Supported	2xx		-	o	o	m*	m*	o	o
Timestamp			o	o	o	o	o	o	o
To	c	r	m	m	m	m	m	m	m
Unsupported	420		-	m	-	m	m	m	m
User-Agent			o	o	o	o	o	o	o
Via	R	amr	m	m	m	m	m	m	m
Via	rc	dr	m	m	m	m	m	m	o
Warning	r		-	o	o	o	o	o	o
WWW-Authenticate	401	ar	-	m	-	m	m	m	m
WWW-Authenticate	407	ar	-	o	-	o	o	o	?

Notes: The "where" column describes the request and response types in which the header field can be used. An empty entry in the "where" column indicates that the header field may be present in all requests and responses. Values in this column are:

R: header field may only appear in requests.

r: header field may only appear in responses.

2xx, 4xx, etc.: A numerical value or range indicates response codes with which the header field can be used.

c: header field is copied from the request to the response.

The "proxy" column describes the operations a proxy may perform on a header field:

a: a proxy can add or concatenate the header field if not present.

m: a proxy can modify an existing header field value.

d: a proxy can delete a header field value.

r: a proxy must be able to read the header field, and thus this header field cannot be encrypted.

The next six columns relate to the presence of a header field in a method:

c: conditional; requirements on the header field depend on the context of the message.

m: the header field is mandatory for client to send and server to understand.

m*: the header field SHOULD be sent, but clients/servers need to be prepared to receive messages without it.

o: the header field is optional for the sender, may be ignored by receiver.

t: the header field SHOULD be sent, but clients/servers need to be prepared to receive messages without that header field. If a stream-based protocol (e.g., TCP) is used as a transport, then the header field MUST be sent.

*: the header field is required if the message body is not empty.

-: the header field is not applicable and must not be included.

?: Not defined or documents are ambiguous.

Source: Based on Tables 2 and 3 from RFC 3261 with additions from later extensions in RFCs.

Allow: when present, lists all the methods supported by the UA sending the message. Absence of this header doesn't mean anything.

Authentication-Info: information, such as a nonce, for mutual authentication with HTTP Digest.

Authorization: carries the credentials of the sending UA. This header is unusual because it can be present multiple times in a message, and they must not be combined by a proxy.

```
Authorization: Digest username="bob",
  realm="biloxi.com",
  nonce="dcd98b7102dd2f0e8b11d0f600bfb0c093",
  uri="sip:bob@biloxi.com",
  qop=auth,
  nc=00000001,
  cnonce="0a4f113b",
  response="6629fae49393a05397450978507c4ef1",
  opaque="5ccc069c403ebaf9f0171e9517f40e41"
```

Call-ID: (i:): may be abbreviated "**i:**" in a message. A large random number, generated cryptographically by the calling UAC. Intended to be globally unique, so the random number is often combined with a separator ($@$, {, }, etc.) and the domain.

```
Call-ID: f81d4fae-7dec-11d0-a765-00a0c91e6bf6@biloxi.com
```

Call-Info: extra information about the sender such as a "purpose = " statement, an "icon" to designate an image related to the call, a "card" in the form of a vCard, or just "info" in free form. Can be dangerous as the "info" could be a URL of a malicious website.

Contact: in an INVITE can be the display name or a URI. The REGISTER request puts the URI of the registering entity in this header. In a 3xx response it's the forwarding location. Either may contain a "q = " and/or "expires" parameter. This header name can be abbreviated m, moved.

```
Contact: "Mr. Watson" <sip:watson@worcester.bell-telephone.com>
     ;q=0.7; expires=3600,
     "Mr. Watson" <mailto:watson@bell-telephone.com> ;q=0.1
```

Content-Disposition: tells the receiver how to handle the message body, using MIME types and additional types defined for SIP. "Session" indicates the body is SDP (the default if this header is missing is Content-Type = - application/sdp); "render" means display to the user and is the default for other Content-Types. Also possible are alert, icon, and any MIME type, which can determine handling.

Content-Encoding: modifies the "media-type" designation with required decoding necessary to render the medium. Compression is a common encodings. Multiple encodings are listed in the order applied, so they can be reversed in order.

Content-Language: short form of a language name: en English, fr French, and so forth.

Content-Length: (l:): decimal number of octets in the message body, 0 or greater. Locates the end of a message when streamed (required for TCP) rather than framed by UDP.

Content-Type: the media type in the message body; required if the body is not empty. These types are registered with and listed by IANA. May have parameters (e.g., character set).

CSeq: a decimal number that can be written as a 32-bit unsigned integer, followed by the SIP method, which is case sensitive. The UAC starts randomly, then increments with each message in a dialog.

Date: required in almost every case. UAs that don't have a clock must not set a date header. Messages received without a Date header should have one added by the next server.

```
Date: Sat, 12 Nov 2011 23:29:00 GMT
```

Error-Info: may include an audio message, a URI for a site that plays an audio recording, or text for a pop-up to display.

```
SIP/2.0 404 The number you have dialed is not in service
Error-Info: <sip:not-in-service-recording@atlanta.com>
```

Expires: time in seconds for the content of the message to remain valid, up to $(2^{32})-1$. For example, the life of a registration or a temporary relocation. Differs from HTTP practice which is to state the absolute date and time of expiration rather than a relative time.

From: (f:): the original sender of the request, whether caller or callee. Display name, URI, and parameters follow the construction of the Contact: header.

In-Reply-To: a Call-ID of a previous message from the UA now being called. Used by call distribution systems (ACDs) to reach the originator of a session. A filter could block calls from UAs you didn't call first.

Max-Forwards: an integer ($0-255$, 70 recommended), the number of hops remaining, decremented by every server; triggers discard at 0. Required in all SIP methods. Can be used like a trace route to troubleshoot loops.

Min-Expires: shortest refresh interval supported by a server, up to $(2^{32})-1$ seconds. A UA that wants a faster refresh will get the error 423 (Interval Too Brief).

MIME-Version: if present, this header claims that the message complies fully with MIME conventions, which is not required of SIP messages in general. Default is 1.0, but 1.1 is used.

Organization: the company or agency sending the message.

Path: records and reports intermediate proxies between a UAC and a Registrar. Used by a proxy to route messages to a UA when one or more proxies must be traversed and those proxies are not identifiable through DNS or SIP messages. Looks like a Record-Route: header. Usually inserted by a proxy, not the UAC. [RFC 3327]

Priority: the importance of promptness, specified by the UAC as "nonurgent," "normal," "urgent," and "emergency" (which should be reserved for threats to life and limb).

Proxy-Authenticate: an authentication challenge from a proxy to a client, required in a 407 (Proxy Authentication Required) response. The form of the challenge dictates the procedure required and what credentials are expected. The UAC (and not the proxy) must provide its credentials in a Proxy-Authorization: header to prevent the proxy from incrementing CSeq, which would cause the UAC to reject the UAS's response (CSeq would not match). A proxy that forks a request to multiple UAs may receive multiple Proxy-Authenticate challenges, which it must aggregate into a single message to the UAC.

Proxy-Authorization: may be included in initial request by UAC. In addition to the caller's credentials, a "realm" parameter in this header addresses it only to proxies in that realm. The header is "consumed" by the proxy server that requested it and does not propagate further.

Proxy-Require: list of option tags (parameters registered by IANA) for SIP extensions that the UAC insists be supported by all proxies in the path. Tags must be from standards-track IETF documents.

Record-Route: inserted by a proxy to force messages from the dialog to pass through it.

```
Record-Route: <sip:server10.biloxi.com;lr>,
              <sip:bigbox3.site3.atlanta.com;lr>
```

Reply-To: a logical URI for return messages, missed calls, and so forth.

Require: list of option tags for SIP extensions that the UAC states are necessary to handle the request. See also Supported, Unsupported, and Proxy-Require.

Retry-After: included by UAS in a response 500 (Server Internal Error) when it receives a new INVITE before sending a final (2xx) response to an earlier INVITE. The value is set randomly to between 0 and 10 seconds. Also appears in other 4xx and 6xx responses to indicate how long a service is expected to be out.

Route: forces a message to pass through the listed proxies.

```
Route: <sip:bigbox3.site3.atlanta.com;lr>,
       <sip:server10.biloxi.com;lr>
```

Server: the software name and version that is providing proxy functions.

Subject: (s:): same as in an email.

Supported: list of option tags for SIP extensions that the sending UA supports. See also Require and Proxy-Require.

Timestamp: when the UAC sent the request to the UAS, format not defined here. HTTP uses only GMT so no time zone is needed when expressed as:

```
Timestamp: Sun Nov 6 08:49:37 1994
```

To: (t:): logical recipient of a request: the callee. May include a "Display-Name:" parameter; must include the unique "Tag" parameter. Tags in To: and From: headers must be globally unique, for which the practical requirement is a cryptographically generated number with at least 32 bits of randomness.

```
To: The Operator <sip:operator@cs.columbia.edu>;tag=287447
t: sip:+12125551212@server.phone2net.com
```

Unsupported: list of option tags for SIP extensions that can't be handled.

User-Agent: software version of the UAC. Can reveal vulnerabilities to hackers.

Via: (v:): names the transport protocol ("UDP," "TCP," "TLS," or "SCTP") and the IP address where a response must be sent. An optional "rport = " attribute gives the L4 port of the public source address of this message, to be used in any response so it will pass the filter of a NAT firewall. Via is required from a UAC in a request. Must contain a "branch" parameter (starting with "z9hG4bK") plus a random number to allow matching responses to the request. RFC 2543 elements don't require a Branch value; for interoperability it is not required by a receiver. Branch values must be globally unique except in messages CANCEL and ACK for non-2xx responses, in which the branch must match the one in the INVITE it cancels or the response it acknowledges.

```
Via: SIP/2.0/UDP erlang.bell-telephone.com:5060;branch=z9hG4bK87asdks7
Via: SIP/2.0/UDP 192.0.2.1:5060 ;received=192.0.2.207;branch=z9hG4bK77asjd
```

Warning: a three-digit warning code, listed in Table 5.9, a host name, and a warning text in a natural language likely to be understood by the user.

WWW-Authenticate: an authentication challenge from a UAS in a response message 401 (Unauthorized). The parameters should indicate the form of credential expected. The UAC should re-originate the request with credentials in an Authorization header, incrementing CSeq. You can always try to log in as Anonymous with no password (" ").

5.3   SESSION DESCRIPTION PROTOCOL

The purpose of SDP is to provide enough information about a session, so that:

- The recipient can decide whether to participate (based on required bandwidth, media formats, etc.).
- If joining a session, the recipient will know where and how to join.
- The recipient will follow a pointer or URI for further information or media sources.

TABLE 5.9   SIP warning header codes (3xx)

Code	Display Text	Meaning
300	Incompatible network protocol	One or more network protocols in the session description are not available.
301	Incompatible network address format	One or more network address formats in the session description are not available.
302	Incompatible transport protocol	One or more transport protocols described in the session description are not available.
303	Incompatible bandwidth units	One or more bandwidth measurement units contained in the session description are not understood.
304	Media type not available	One or more media types in the session description are not available.
305	Incompatible media format	One or more media formats in the session description are not available.
306	Attribute not understood	One or more of the media attributes in the session description are not supported.
307	Session description parameter not understood	A parameter other than those listed above was not understood.
330	Multicast not available	The site where the user is located does not support multicast.
331	Unicast not available	The site where the user is located does not support unicast communication (usually due to the presence of a firewall).
370	Insufficient bandwidth	The bandwidth specified in the session description or defined by the media exceeds that known to be available.
399	Miscellaneous warning	Can include arbitrary information to be presented to a human user or logged. A system receiving this warning must not take any automated action.

Note: Warn-Code values are registered with IANA.

When a UAC initiates a dialog, it typically includes a SDP block in the body of the INV message. An SDP session description requires:

```
Session name and purpose
Time(s) the session is active
The media comprising the session
Information needed to receive media (addresses, ports, formats, etc.)
```

Multiple media and versions may be listed; the UAS can then see if it supports at least one of them and therefore can handle the requested session. The UAS may respond with its own list of alternatives, for example, its audio and video codecs. The overlap in the lists is what they are capable of using in a session (Figure 5.6). The choice of codec or media depends on the preference rating each UA attaches to each option in its own list as a quality or "q= "

value between 0 and 1. The higher the q value, the greater is the preference. Combining the ratings from both sides ranks the options, the highest value wins and that option is used.

There is a shortcoming in SDP at this writing: it allows only one IP address for all sessions. This was appropriate for IPv4, but with IPv6 a device may have many IP addresses. Also, as carriers and enterprise backbones migrate from v4 to v6, they may run both stacks. An end point may find that it can reach signaling servers via IPv6, but the other end point is limited to v4. A situation like this could require different versions of IP for control and media streams. The IETF is working on solutions to make the IP version migration more interoperable.

The UAC need not include SDP information in the INV message; it can be blank. In that case the UAS must supply its SDP block in the first response that is "delivered with a reliable method." That may be a Trying or an OK response. The UAC then delivers its SDP block in the next message, the ACK.

Some users may lack resources necessary to participate in a session or may not be interested. Additional information that fits in a description is a list of categories or classifications of the content that the user can filter on and also:

```
Information about the bandwidth to be used by the session
Contact information for the person responsible for the session
```

SDP is important to SIP, but SDP also participates in Session Announcement Protocol (SAP) to advertise multicast media events. SDP has a registered MIME type:

```
media type = "application/sdp"
```

Fortunately for most users, SDP lies buried deep in application software and appears to them mostly through warnings and options. For example, client software must ask a user for permission before sending a media stream or following a link. Error messages should raise an alarm or warn the user.

If you really want to know about SDP, the field names and attribute names use only the US-ASCII subset of UTF-8, but textual fields and attribute values may use the full ISO 10646 character set. When SDP information travels in the clear, there is some risk that it might be useful to a hacker who could join a broadcast or glean useful facts from the session parameters. For privacy, you can encrypt the SDP body. To a recipient who lacks the key, the message appears malformed and is discarded.

SDP messages consist of a number of lines of text of the form:

```
<type>=<value> no white spaces allowed
```

where <type> is exactly one case-significant character and <value> is structured text whose format depends on <type>. The formats are defined formally in ABNF (see sidebar).

The message body begins with a session-level section whose first line is a "v = " version number of SDP. There follow zero or more media-level sections. Here is where the options appear. Each starts with an "m = " line and contains specific attributes in "a = " lines and others. For example:

```
v=0
o=jdoe 2890844526 2890842807 IN IP4 10.47.16.5
s=SDP Seminar
i=A Seminar on the session description protocol
u=http://www.example.com/seminars/sdp.pdf
e=j.doe@example.com (Jane Doe)
c=IN IP4 224.2.17.12/127
t=2873397496 2873404696
```

IN SHORT: ABNF

Augmented Backus–Naur Form is a notation used to define syntax that describes precisely how information is expressed. Precision eliminates ambiguity to allow computers to parse the meaning from a text string. ABNF does this by imposing rules for definitions:

Characters are US-ASCII, but case-insensitive to printable representations. Thus a and A are equivalent. A rule is given by its name, at least one space, an equal sign, a space, and the expression.

```
rule   = "abc" matches abc, Abc, aBc, abC, ABc, aBC, AbC, ABC
```

Certain basic rules are in uppercase, such as SP, HTAB, CRLF, DIGIT, ALPHA.

To specify case, the character is escaped, with a specific indication of the number base: d = decimal, x = hexadecimal, b = binary.

```
rulename  = %d97 %d98 %d99 CRLF  matches only lowercase abc
CR        = %d13
CR        = %x0D
LF        = %d10
```

A period separates characters in a concatenated string.

```
CRLF   = %d13.10
```

Elements separated by a forward slash ("/") are alternatives, like the vertical bar "|" in regular expressions.

```
foo / bard   matches either foo or bard
```

```
a=recvonly
m=audio 49170 RTP/AVP 0
m=video 51372 RTP/AVP 99
a=rtpmap:99 h263-1998/90000
```

All session description lines of the headers must appear in exactly the order listed below, to make parsing easier and to reveal errors in creating the message:

- Session
- Time
- Media

A hyphen ("-") indicates a range of alternative values.

```
DIGIT  = %x30-39 is equivalent to
DIGIT  = "0"/"1"/"2"/"3"/"4"/"5"/"6" /"7"/"8"/"9"
```

Elements enclosed in parentheses are treated as a single element whose contents are strictly ordered.

The "*" preceding an element indicates repetition of that element at least a and at most b times. It is used to specify a number of digits in a field, a minimum length of password, and so forth.

```
 <a>*<b>element
*<element> allows any number, including zero; 1*<element>
requires at least one; 3*3<element> allows exactly 3; and
1*2<element> allows one or two.
 <n>element is equivalent to <n>*<n>element: exactly <n>
occurrences.
```

Square brackets enclose an optional element sequence:

```
[foo bar]  is equivalent to *1(foo bar).
```

Combining concatenation with alternatives gets messy. See RFC 4234, *Augmented BNF for Syntax Specifications: ABNF*. This example of ABNF applied to field descriptions comes from SDP:

```
bandwidth-fields = *(%x62 "=" bwtype ":" bandwidth CRLF)
     time-fields = 1*( %x74 "=" start-time SP stop-time
                 *(CRLF repeat-fields) CRLF)
                 [zone-adjustments CRLF]
```

Those marked "*" are optional, which may be omitted but not placed elsewhere in the message.

Attributes of a session entity apply to all media sections unless overridden by another attribute in a media section. There the attribute applies only to the medium.

```
v= (protocol version)
o= (originator and session identifier)
s= (session name)
i=* (session information)
u=* (URI of description)
e=* (email address)
p=* (phone number)
c=* (connection information - not required if included in all
  media)
b=* (zero or more bandwidth information lines)
```

One or more time descriptions apply to each session. Times for sessions to start and end are expressed in the format of Network Time Protocol (NTP), that is, seconds since 1 Jan 1900 is a decimal version of a 64-bit number. Shades of Y2K, this number will roll over in 2036. Time is represented consistently everywhere, including local adjustment for daylight savings time (DST).

```
t= (time the session is active)  starttime stoptime [in NTP
  format]
r=* (zero or more repeat times)
z=* (time zone adjustments)
k=* (encryption key)
```

Not recommended except over a secure channel.

```
a=* (zero or more session attribute lines)
```

Zero or more media descriptions follow; if present, they override session parameters:

```
m= (media name and transport address)
i=* (media title)
c=* (connection information - optional if included at session
  level)
b=* (zero or more bandwidth information lines)
k=* (encryption key)
a=* (zero or more media attribute lines)
```

Media types "audio," "video," "text," "application," and "message" are recommended. Former values of "control" and "data" are deprecated.

The connection, c =, may include a protocol indicator. Registered protocols originally were RTP with a defined AV profile ("RTP/AVP"), secure RTP (sRTP), and UDP. Within a protocol description, there may be a number code to indicate the format. Address type may be indicated as "IP4" or "IP6."

An SDP parser ignores any session description it doesn't understand. The attribute mechanism ("a = " line) is where vendors can extend SDP. Some attributes have a defined meaning, registered with IANA, but vendors can use others for an application, a medium, or any other basis.

RFC 4566, *Session Description Protocol*, defines the formats in ABNF notation and the originally registered field names ("media", "proto", "fmt", "att-field", "bwtype", "nettype", and "addrtype"). The media types are limited to "audio," "video," "text," "application," and "message" with a high barrier set to discourage new types.

5.4 MEDIA GATEWAY CONTROL PROTOCOL

The MEdia GAteway COntrol (MEGACO) protocol evolved from earlier versions developed separately by the ITU and the IETF (also called Media Gateway Control Protocol but abbreviated MGCP; RFC 3435). They merged into a joint development program with the results published under both numbering systems as H.248 and RFC 3525. ITU has now taken responsibility for Megaco and published two later versions as H.248.1 version 2 (in 2002) and version 3 (in 2005). See RFC 5152 for a history of the changes that made RFC 3525 obsolete.

Extensions to H.248.1 can appear as new RFCs and as additional Recommendations numbered H.248.nn. As of this writing, later documents have raised nn to more than 60.

Much of the detail work for Megaco occurs below an Application Programming Interface (API) on third-party code. It is a protocol stack that equipment manufacturers routinely buy rather than build as a way to get products to market faster.

Media gateway (MGW) products may claim compliance with any version of H.248.1. Users should determine if the feature set of the version claimed by a device being considered includes all the features needed. It's better to have it in writing, from the appropriate data sheet or release notes, rather than oral assurances.

5.4.1 MGW Functions

Media gateways sit between different types of networks. The logical gateway between networks provides multiple functions, which may reside physically in different servers or specialized hardware platforms. A MGW translates the information on one side into the formats required on the other side (Figure 5.7). Thus an IP gateway will apply its internal signaling and media capabilities to convert between:

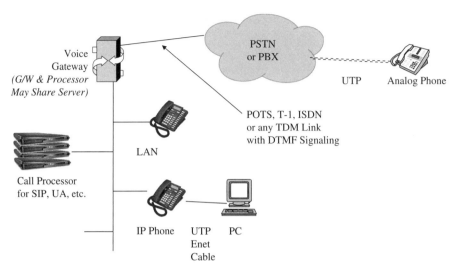

FIGURE 5.7 Megaco protocol, which allow a VoIP call control server to treat PSTN connections like IP phones, converting between packets and continuous streams.

- SIP or H.323 signaling on the IP network, and for examples:
  1. Common channel signaling on an ISDN line.
  2. DTMF.
  3. On/off-hook conditions for a POTS line.
  4. Robbed-bit Channel Associated Signaling (CAS) on a T-1 or E-1 trunk.
- IP packets of UC information on a LAN and continuous streams of voice or video bits in a digital channel or on an analog line.

Larger MGWs may be partitioned into multiple virtual gateways, each with designated capacity and access to certain physical ports. Each virtual segment can register with a different MGC or a process thread. A CPU core in a MGC might be dedicated to a logical subset of a MGW. Each MGC establishes a separate stream with the MGW, but only one stream per partition. To prevent conflicts, only one MGC (the master) has write permission to the physical terminations; all other MGCs control contexts for a static set of assigned terminations.

The Megaco protocol relationship is between a master (MGC) and one or more slaves (MGWs). The MGC can deal with any number of MGWs but each logical or physical MGW registers with only one MGC.

For higher availability of gateways, they and MGCs have automatic failover modes. A MGW may be configured with the addresses of multiple MGCs. If the active MGC fails to respond, the MGW will attempt to register with the next MGC on its list. The MGC typically handles multiple MGWs. If one MGW is lost, the MGC routes traffic to another MGW (see Table 5.10).

TABLE 5.10   Media gateway services between IP phone system and PSTN

IP Phone Places Call		Action Between MGW and Trunks		
Transport Function	MGW/Megaco Action	Analog POTS	ISDN Local Loop	T-1 Local Loop
Originate a call	MGC commands MGW to seize trunk with Add command	MGW port goes off hook	Signaling message on D channel	Change in robbed bits (CAS)
Dial phone number	Number in signal H.248.1 packet	DTMF tones	D channel message, DTMF in B channel	DTMF in voice channel
Voice transport	Voice in packets passed to digital line or changed to analog	Analog audio	PCM in DS-0 digital channel	PCM in DS-0 digital channel
Clear call	Command to remove all terminations from context	MGW Port goes on hook	Signal in D channel	Robbed bits (CAS) return to idle

Receive Call from PSTN/PBX		Action Between MGW and Trunks		
Transport Function	VoIP (SIP) [H.323]	POTS Analog Trunk	ISDN Local Loop	T-1 Local Loop
Announce a call	MGW notifies MGC	Ringing voltage from switch	D-channel message	Change in robbed bits (CAS)
DID data	MGW collects all digits, sends update to GK	DTMF or modem data between rings	D-channel message from switch	DTMF or modem data between rings
Answer supervision	IP phone sends OK MGC tells MGW to Connect	MGW port goes off-hook	D-channel message from MGW	MGW changes robbed bits (CAS)
Voice transport	May pass through, en/decode, transcode	Analog audio	PCM in DS-0 channel	PCM in DS-0 channel
Clear call	Command to remove all terminations from context	Switch goes to idle when MGW port is on-hook	D-channel message from either side	Either side returns robbed bits (CAS) to idle

Note: Topology assumed is MGW with a TDM trunk to the PSTN and an IP phone on a LAN.

The MGW, the slave device, performs almost nothing on its own. It deals with only the simplest procedures, then reports them to the MGC for processing as events. One thing the MGW does well is collect dialed digits from a PSTN interface or analog telephone and compare them to a dial plan configured ahead of time or selected by the MGC. When the MGW recognizes sufficient numbers, it will notify the MGC and await instructions. The MGW may recognize Dialed Number Indication, or the Direct Inward Dialing number, received from the PSTN. A Notify message will carry the target address or extension number.

When linking a TDM PBX to a UC system, the MGW would recognize patterns such as a leading 8 indicating an outside call, a local area code plus seven digits, or 4 digits for an on-site extension. The event notification from the MGW would let the MGC continue the call on the IP side.

On "outbound" calls from LAN to PSTN, the call MGC will convert the caller's SIP address, if a URL, to an E.164 phone number that the MGW can give to the PSTN in the form of dialed digits. The called phone number may also be input manually by the caller on the keys of an IP phone, which would capture the dial string and send an INV to its proxy.

5.4.2 MGW Connection Model

Before describing the protocol procedures for Megaco, it is necessary to see how a MGW looks internally, to understand the actions of Megaco commands. This MGW model is a logical construct and does not imply any constraints on hardware or firmware. Figure 5.8 depicts its internal structure modeled for Megaco.

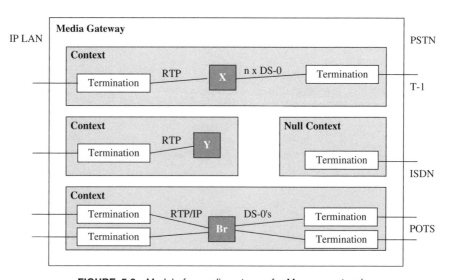

FIGURE 5.8 Model of a media gateway for Megaco protocol.

A MGW connects to external networks on physical interfaces represented by terminations. These are logical entities that originate or receive media traffic or signaling. A termination may represent a channel permanently provisioned on a physical port, for example, a POTS line from the PSTN, or a logical UDP connection to the IP network. A media gateway controller (MGC) creates a connection on the MGW by sending an Add command that puts terminations into a "context" or logical association with each other. Functions like transcoding or bridging are added to a context as needed.

The MGW takes down a connection by applying a "subtract" process to each termination. Removing the last termination deletes the context.

A context may also receive a "modify" command to adjust parameters during a call, such as changing the bit rate of a video codec. Which parameter the MGC wants the MGW to change is identified in a Descriptor (Table 5.12).

Terminations can respond to stimuli such as DTMF tones or signaling messages. The result is a "notify" message sent to the MGC. When configured for redundancy, MGWs use "Notify" to advise the MGC when switching to a backup unit and when a failed unit returns to service.

The nature of the terminations and how they are joined in a context determines what the MGW needs to do. An ISDN Termination will recognize Q.931 signaling messages on the D channel. An Ethernet interface that deals with VoIP will recognize SIP messages or H.323 signaling. When more than two terminations are added to a context, the result may be a conference call or a broadcast, depending on how the terminations and the context itself are defined.

A topology descriptor (TD) applied to a context controls how information flows in and out of terminations. Full duplex transmission would be normal for a phone conversation, but a video conference might be managed to designate the termination of the current speaker as the source of the video stream sent out on all other terminations. That is, the participants see only the speaker of the moment. By default, no TD connects all terminations in full duplex mode. Quite complex arrangements of stream flows are possible.

For each medium in a context the MGC creates a locally unique StreamID, an identifier number. The stream may have attributes.

Some common actions follow, with descriptions of how they manifest in a MGW:

- **Hold:** a termination is placed alone into a new context, a null context.
- **Transfer:** a termination is "moved" to another context.
- **Conference:** a specialized bridge termination (BR) for conferencing is added to a context.
- **Multiplex:** termination entities for multiplexing, inverse multiplexing, and load sharing exist in some media gateways. An example is to create an $N \times 64$ kbit/s channel on a T-1 link for some media stream.

The ITU has defined "packages" for Megaco that group certain properties and behaviors such as tone generation and line supervision. These are related feature sets with clearly defined scope and purpose. If both the MGW and the MGC implement a specific package, they are required to interoperate for all the features in the set.

5.4.3 Megaco Procedures

Information between the MGC and MGW passes over an IP network using the Megaco protocol. An earlier version called MGCP and proprietary versions of signaling exist but seem to be declining in popularity compared to the standard, H.248.1.

Messages between the MGC and MGW (from H.248.1) process a call in small increments. The following is an example of message exchanges for a call between two residential MGWs. That is, one legacy phone calls another over an IP network with an MGW at each side.

Before the call, two analog phones are assigned (by configuration or geography) to the different MGWs, -a and -b. When an MGW first connects to an IP network, it finds and registers with a media gateway controller (MGC). To simplify this example of Megaco, the MGC will take on the functions of any SIP proxies and registrars. The dial plans are installed by administrative action.

- Each MGW registers with the same MGC using a ServiceChange command.
- The MGC response to each MGW confirms the configuration profile, IP addresses, and services supported.
- The MGC assigns TerminationIDs for each interface in each MGW and assigns each termination to a NULL context, a context with only one termination.
- Each MGW responds to indicate the command was accepted, while monitoring terminations for activity or events.
- MGW-a detects an off-hook condition when the caller picks up the phone and informs the MGC with a Notify command.
- MGC acknowledges the Notify.
- MGC sends a Modify command to MGW-a for the termination that detected off-hook, so it plays a dial tone to the caller, watches for an on-hook event, and awaits dialed digits.
- MGW-a acknowledges the Modify command.
- MGW-a accumulates dialed digits, turning off dial tone on receipt of the first digit. The previously loaded dial plan informs the MGW how to process dialed digits and when enough have been received to notify the MGC. At that point MGW-a sends the dialed string in a Notify message to the MGC.
- MGC acknowledges the Notify.
- After MGC analyzes the dialed digits, it understands where the call goes, in this case to MGW-b. MGC adds the TDM termination from the caller's

phone and an RTP termination to a new context in MGW-a. The phone's mode is set to RecvOnly because the description of the far end isn't yet available.

- MGW-a acknowledges the new context, assigning a local IP address and UDP port as well as an RTP port for the connection.
- MGC puts the called termination in a new context with an RTP Stream and tells MGW-b to ring the phone.
- MGW-b acknowledges the configuration with an SDP block that contains a port number for the RTP Stream, identifiers, and the audio profile.
- MGC forwards the IP and port of MGW-b to MGW-a, specifying the context to be updated.
- The end-to-end connection on RTP triggers a ringback tone to the caller.
- When the called phone goes off-hook, MGW-b notifies MGC.
- MGC uses a Modify command to turn off ringing at MGW-b.
- MGC change MGW-a to send/receive mode with a Modify command.
- MGW-a acknowledges the Modify.
- The two MGWs are now able to communicate. During the call the MGC may Audit either gateway to examine its configuration or performance.
- When either MGW goes on-hook, it notifies the MGC.
- MGC ACKs the Notify.
- MGC sends Subtract commands to both MGWs to empty the contexts, which clears the call. Terminations return to Null contexts.

If the Null context is configured for terminations to watch for incoming alerts, then no further messages from the MGC are needed to return to a normal idle condition.

Megaco functions with a short list of commands, shown in Table 5.11.

Commands contain parameters, which are attached to descriptors. Text format looks like

```
DescriptorName=<someID>{parm=value, parm=value...}
```

where a parameters may be:

- **Fully specified:** a single value to use for the specified parameter.
- **Underspecified:** the parameter value CHOOSE allows the command responder to pick a value it can support (the DS-0 for an outbound call on a PRI).
- **Overspecified:** a parameter list of potential values, in the order of preference; the responder chooses a value and returns it to the initiator.

The MGC attaches event descriptors to a termination as it is placed into a context. The descriptors tell the MGW how to handle events that occur on that termination. Table 5.12 explains the descriptors briefly.

TABLE 5.11   Megaco command messages

Command	Function
Add:	Puts a termination into a context; creates a context when adding a first termination.
Modify:	Changes properties, events, or signals of a termination.
Subtract:	Disconnects a termination from its context; deletes a context when subtracting the last termination.
Move:	Moves an existing termination from one context to another.
AuditValue:	Returns the current properties, events, state of signals, and statistics of terminations.
AuditCapabilities:	Returns all possible values for allowed termination properties.
Notify:	Allows the media gateway to inform the media gateway controller of events.
ServiceChange:	MGW registers with the MGC: MGW notifies the MGC of restart or that one or more terminations are going out of service or have returned to service; MGC announces a handover to the MGW; the MGC instructs the MGW to put one or more terminations in or out of service; the active MGC can hand off the MGW to another MGC.

TABLE 5.12   Descriptors in Megaco messages

Name	Function
Mux	Describes the sources of media streams and how they are multiplexed into a session.
Media	List of media streams and their specifications.
TerminationState	Properties of a receiver, (e.g., a list of packages supported) not specific to a stream.
Stream	The local, remote, and control descriptors for a stream.
Local	Properties of media streams received by the MGW.
Remote	Properties of media streams sent by the MGW.
LocalControl	Properties, as in packages, used between the MGW and its MGC.
Events	Instructions to MGW on events to watch for and what to do for each.
EventBuffer	Events to watch for when event buffering is on.
Signals	Instructions for MGW to apply to terminations.
Audit	Request for information.
Packages	In Audit message, obtains list of packages for each termination.
DigitMap	List of multiple events (e.g., received dialed digits) to match and handle together.
ServiceChange	Requests what and why of changes in MGW.
ObservedEvents	In Audit or Notify request, reports events.
Statistics	Requests data on terminations, in Subtract or Audit message.
Topology	Specifies directions of flows among terminations in a context.
ContextAttribute	Usually a function of a package, applies to entire context.
Error	Delivers a cause number in a Notify or any response message.
Modem	Specifies modem type in version 1 only. Deprecated—do not use.

Dial Plan:

Dialed Digits	Description
0	Local operator
00	Long distance operator
Xxxx	Local extension (X = 1 to 7)
8xxxxxxx	Local number
#xxxxxxx	Off-site number
*xx	Star services
91xxxxxxxxxx	Long-distance number
0911 + ≤ 15 digits	International access

DigitMap:
(0| 00|[1-7]xxx|8xxxxxxx|Fxxxxxxx|Exx|91xxxxxxxxxx|9011x.)

FIGURE 5.9   Patterns in a DigitMap derived from a dial plan.

To interpret dialed digits it receives, the MGW is configured with one or more DigitMaps, a list of number patterns that the MGW should recognize and report as a single event to the MGC with a Notify message. These are dial plans that allow the MGW to process numbers in order to direct them or modify (filter) them. The DigitMap applies to outbound calls as well as inbound, allowing a system to block certain calls from specified classes of users. For example, phones in public spaces can be limited to local calls.

A MGW between a legacy phone and an IP network would recognize 8 in the first position, the sequence 911, zero, and possibly other patterns that fit abbreviated dial plans within the organization. A call by an emergency response team can get priority over normal calls. Figure 5.9 compares a sample dial plan with the DigitMap derived from it. The occurrence of these patterns would constitute and event and generate a Notify message to the MGC. The MGC then instructs the MGW how to deal with the event.

Signals handled by the MGW include received dial digits, but the MGW can also detect error tones, on/off-hook conditions, and other events. The MGC indicates which termination in a context is to send a signal.

5.4.4   Megaco Details

Reflecting its dual parenthood, Megaco has two modes, each with a default port number: 2944 for text-encoded operation or 2945 for binary-encoded operation. Responses go to the same port from which the command came. There are no default ports for responses. Reduction in bandwidth needed for messages is possible with the binary coding because a 64-character text string is equivalent to a 2-octet binary number.

Binary encoding of H.248.1 messages optimizes machine processing. When examining these transmissions on the line, the protocol analyzer needs to apply the ITU encoding scheme (two hex characters per byte) and recognize the order

TABLE 5.13   Encoding conventions for Megaco messages

Binary Octet Bit Pattern	Hexadecimal Coding (0xFF)
00011011	D8
11100100	27
10000011 10100010 11001000 00001001	C1451390

of bit transmission, (most-significant byte then least significant bit goes first). Examples from H.248.1 in Table 5.13 best describes it.

Within a binary octet, the LSB is written to the left, the MSB to the right, which is the order of transmission on a network interface. Hex digits each represent half an octet, but they are written in "decimal" order with the most significant character on the left. Hence the written orders of the hex and binary codings are not the same.

Text encoding may use full words and indentation, or abbreviations without indents. Vendors also have nonstandard message encoding methods for faster internal processing.

Since ITU took over Megaco, the enhancements have leaned toward carrier-related tools such as additional ways to change services on a MGW and better reporting of availability of hardware components like ports and processors. ITU also organized many of the features into sets, called packages. There are 15 basic packages including DTMF detection or generation, answer supervision of analog lines, and so forth. Like the SIP Connect document, the packages represent progress toward better interoperability.

As an indication of a need for profiles as well as standards, AudioCodes offers six ways to carry DTMF signals received from the TDM side. On the packet side, in addition to transmitting tones as encoded voice, there are proprietary methods (Nortel, Cisco) and multiple standards ("DTMF relay" per RFCs and Internet drafts) as well as a national standard (in Korea). Configuring the call control server and media gateway controller of course must match the MGW.

Most actions of the MGW either respond to a stimulus on a termination (incoming signaling information), with a report (Notify) to the media gateway controller (MGC), or respond to a command from the MGC. Additional functions include the "audit" command from the MGC that return properties and configurations set in the MGW. The MGW may use the ChangeService message to register with the MGC when it comes online or to advise that some hardware is leaving or entering service. The MGC uses that message as a command to control the up/down status of terminations or the entire MGW.

There is little difference in basic results if an IP phone communicates with its call controller using SIP, H.323, or Megaco. Some IP phone have been designed exclusively for MGCP. They treat the handset as the POTS trunk, with analog interfaces. The MGC treats the phone as a single-port MGW.

The function of an MGC is a logical subset of the call control process and may be part of a larger IP PBX application. The information flows between a

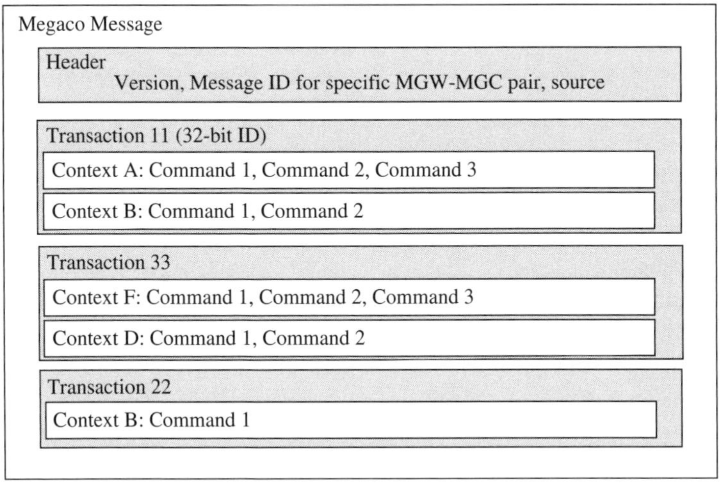

```
Megaco Message

  Header
              Version, Message ID for specific MGW-MGC pair, source

  Transaction 11 (32-bit ID)
    Context A: Command 1, Command 2, Command 3
    Context B: Command 1, Command 2

  Transaction 33
    Context F: Command 1, Command 2, Command 3
    Context D: Command 1, Command 2

  Transaction 22
    Context B: Command 1
```

FIGURE 5.10 Megaco transaction with many contexts and multiple commands for each context. A message may carry multiple transactions.

MGW and MGC group into "transactions." A transaction is a list of commands addressed to one or more contexts (Figure 5.10). A hierarchy within the transaction groups all the commands for each context. Wild cards are allowed when identifying contexts or terminations so a single command can affect many resources in a MGW. Such a command would prepare a multiport module for removal from a chassis, or configure a new module across all its ports.

As MGWs grew larger, signaling transactions grew too and messages exceeded the maximum payload size of UDP. Megaco's version 3 added a segmentation capability to fit this case. MGW equipment probably needs to be configured for segmentation, but it may become active automatically when UDP transport is selected. When Megaco transport is TCP, segmentation is not wanted as that protocol inherently segments and reassembles payloads when needed.

In text format, a transaction looks like:

```
TransactionRequest(TransactionID {
ContextID {Command . . . Command},
  . . .
ContextID {Command . . . Command}})
```

The response uses the same ID numbers:

```
TransactionReply(TransactionID {
ContextID {Response . . . Response},
  . . .
ContextID {Response . . . Response}})
```

Each transaction has a 32-bit ID unique to the MGC. The ContextID is unique for the MGW. A MessageID applied to all messages between a MGW and its MGC remains constant for the duration of the association between the two entities.

Like TCP transmission windows, Megaco transactions are acknowledged when successful. A field in a response message contains the numbers or ranges of transactions to confirm. The grouping of transactions in requests and responses is independent; the message is just transport and does not organize transactions.

The originator of a transaction normally examines the acknowledged ID in a response to prevent executing a command more than once. When a MGW starts up or resumes operation, it performs all actions requested by the MGC without regard to the ID number, just to get started. Also, since the MGW assigns ContextIDs, the MGC command that creates a new context cannot have an ID (the field is set to CHOOSE, leaving it to the MGW).

An MGW that cannot execute a command promptly, but is still processing it, will send a TransactionPending message at intervals controlled by a configurable timer. A failed command halts a transaction.

For reliability, TCP often provides the transport layer. A MGW may implement TCP or an application-level framing (ALF) method to ensure accurate delivery over UDP. The MGC is required to implement both, though TCP appears to dominate deployments. ALF violates the principle that communications functions should be handled by the lowest layer possible. TCP does the job and simplifies the application layer.

Megaco draws on some ISO standards to define how a message like a transaction is carried on a stream-oriented protocol like TCP. The answer is TPKT, described in RFC 1006, which adds a header and a protocol data unit (PDU) in the payload of a TCP packet (Figure 5.11). This form satisfies the need for the higher layer TP0 (transmission protocol zero, an ISO standard) for a network service data unit (NSDU) to indicate a transaction.

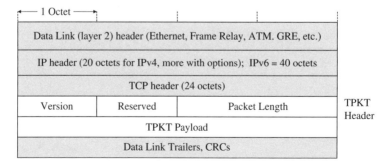

FIGURE 5.11 Megaco transport over a TCP connection involves a TPKT header and PDU.

For security, IPsec is strongly recommended, either the Authentication Header (AH) or Encapsulating Security Payload (ESP), for the link between the MGW and MGC. This is good advice even for paths that lie entirely on an internal network, to prevent malware picking up valuable information from the signaling messages.

5.4.5 Signaling Conversion

Inter-network gateways offered in the late "20 oughts" could handle a wide range of signaling formats, both IP and TDM. Because of the de facto distribution of relatively new hardware at the time, gateways had to be prepared on the LAN side for Cisco's Skinny protocol, a version of SIP, as well as standard formats such as SIP and H.323.

On the TDM side, gateways faced a mix of T-1 interfaces with robbed bit signaling, ISDN primary rate signaling on the D channel, and various PBX interfaces from Q.931 and Q.sig to proprietary PBX formats and DPNSS. Where the PSTN access is analog service, the gateway must support at least loop-start signaling (FXO), and possibly ground-start interfaces. E&M signaling probably won't be required but is available if needed, for example, on adapters to carry land mobile radio over IP.

Software processes typically handle signaling, as the number of options calls for considerable flexibility in configuration. For example, a T-1 line could have 2-bit or 4-bit signaling, regular or extended superframing, and one of at least two options for suppressing long strings of zeros in the data. The options for ISDN are far greater.

Internationally, signaling conversion must deal with the R1 and R2 analog interfaces, channel associated signaling on E-1's, and the E-1 version of the ISDN PRI.

Various vendors offer more specialized gateways to accommodate legacy equipment besides phones. There are gateways specifically tailored to interface to analog voice mail system and fax machines (more on fax elsewhere in this book).

In a bit of a reverse impersonation, several models of IP phone were designed to operate on the Megaco protocol rather than SIP or H.323. These phones work in almost all networks based on any of three main signaling protocols (SIP, H.323, and Skinny) because the call controller would use Megaco rather than those protocols to handle the phone. Not a bad idea for test phones if there is any doubt about the final choice of signaling protocol.

5.4.6 Voice Transcoding

The TDM side, whether the PSTN or a legacy PBX, almost always uses PCM encoding in a DS-0 channel. The LAN side often uses PCM but, depending on the location, availability of bandwidth, and concern about voice sound quality the encoding, could also be compressed or "expanded" to "high-definition voice" of 7 kHz audio bandwidth.

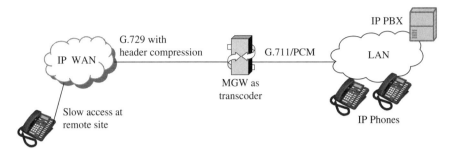

FIGURE 5.12 Transcoding voice channels in a media gateway.

Most of the commonly used voice compression algorithms start from PCM. Receiving that format from the public network simplifies the job of the MGW when it needs to deliver a compressed format on the LAN side—but not by much (Figure 5.12). The analog to PCM encoding is a hardware process, in an inexpensive merchant chip or part of a custom IC. The latency is very low.

Microsoft, for one vendor, uses proprietary codecs in its phones and softphones. This means they require a separate server to transcode between that format of VoIP on the LAN and the PCM-based VoIP at the media gateway. The process introduces latency and requires another server.

When one IP phone calls another, the proxy or control server may recognize that the two phones don't have a common coder, so they can't connect directly. If the network is provisioned with a capable media gateway, and if it and the call controller are configured appropriately, the controller can send an INVITE to the MGW in which the SDP block describes the capabilities of both phones.*

```
m=audio 20000 RTP/AVP 0
c=IN IP4 A.example.com
m=audio 40000 RTP/AVP 18
c=IN IP4 B.example.com
```

If the MGW can transcode for them, it sends separate INV messages to each phone, on separate sessions to set up two calls, and acts as a proxy between them.

5.5 H.323

The first version of H.323 in 1996 was for video terminals communicating over LANs that don't guarantee quality of service. Voice and other components of multimedia came later. Back then, there were many L3 protocols, including NetWare's IPX and Banyan Vines, which is why H.323 was not specific to IP.

---

*Source: AudioCodes manual.

This section looks at the 7th edition of H.323 from 2009. That edition lists 58 other ITU Recommendations that define related protocols or their extensions. The scope of this version includes audio and video codecs, three types of control connections, and the multiplexing function that places the media and control sessions on the packet network.

As recently as 2003 major telephony vendors offered H.323 systems as the only packet-based solution for voice. Within a few years of that time there was almost no telephone product that didn't claim SIP capability.

Recognizing market trends that point to SIP as the VHS of the call-control competition, we will limit our discussion of H.323, the apparent Betamax of voice protocols. The goal here is to cover the key concepts, define terms, and show how the Recommendations operate. The assumption is that new deployments will almost always choose SIP.

5.5.1 H.323 Architecture

Telephone companies traditionally assume that bandwidth is scarce and expensive—usually true to some extent. In response to these conditions the design of control signaling for the PSTN, SS7, and ISDN produced very compact formats. Many fields for different types of information are small, just a few bits each (similar to the 1-bit fields in the IPv4 header, which was standardized in 1983 before broadband became available). The resulting messages, defined by the binary values, are not easily read by humans, but on a datascope or protocol analyzer they are decoded for the benefit of technicians. The later versions of ITU protocols like H.248.1 use some human-readable syntaxes, similar to SDP.

In addition to terminals, an H.323 system will likely include the components of the network similar to Figure 5.13: terminals, gateways, gatekeepers,

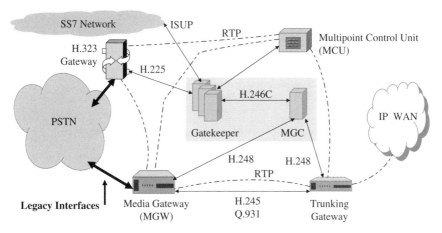

FIGURE 5.13 H.323 architecture, with logical components and protocols. Paths for media sessions are dotted, signal paths are solid. The shaded box encloses the gatekeeper and MGC to indicate that they can be functions of the same device. Registration is not shown.

multipoint controllers, multipoint processors, and multipoint control units. The last three are logical functions often found in one device.

The gatekeeper, terminals, and MCU connect exclusively to the IP network. The gateway connects to the H.323 network and to another network. Most often the other network is the PSTN or a circuit-switched PBX. Many gateways will also link two IP networks where the needed function is to transcode voice (e.g., between G.711/PCM and G.729/CELP) or signaling (between SIP and H.323, Q.sig, or a proprietary format).

The term "H.323" commonly includes all of these components unless the context indicates otherwise. Components communicate in streams, classified as:

- **Audio:** digitized and encoded speech. Each audio stream has an audio control signal.
- **Video:** digitally coded motion video, not still images. Each video stream has a video control signal.
- **Data:** computer files such as still pictures, facsimile images, and documents.
- **Communications control:** signals to exchange capabilities, open and close logical channels, manage mode, and so forth.
- **Call control:** signals to establish and disconnect calls, and so forth.

These streams are packetized according to Recommendation H.225, as separate streams rather than combined into one stream of packets. Each stream can receive a tailored QoS, routing, or other appropriate treatment.

Typically, telco standards in the form of ITU Recommendations define every possible combination of bits and assign them for a specific purpose or "reserve" them for future use. There are no "experimental" fields—reserved bits usually must be set to zero by the sender and ignored by the recipient. Option fields and headers must be standardized, not defined by equipment makers or end users. These restrictions help make the ITU protocols well behaved and predictable—good for interoperability between vendors and networks. The cost is slower introductions of new features.

Like other VoIP environments, H.323 carries management signaling in TCP to ensure accurate delivery while voice and video information travels in UDP to reduce computing overhead. Data streams for documents and computer files also use TCP or a "T.120" connection (Table 5.14).

The registration, admission, and status (RAS) messages don't control calls directly, as do H.225 messages such as Setup. When controlling a call, H.225 travels on reliable TCP. RAS may use UDP.

H.323 is very flexible, with many configurable parameters. For example, a video call between two parties may use the common image format (CIF,

TABLE 5.14   Relationships of protocols in H.323 systems

Data App. or Confer- encing	MGW Control	Session Configur- ation	RAS (Can Tunnel Call Setup)	Call Setup– Clear (Can Tunnel H.245)	Stream Control and Monitoring	Audio and Video Streams
T.120	H.248	H.245	H.225		RTCP	RTP, sRTP
TCP				UDP		
Network layer, IP packets						
Data link layer						
Physical layer						

Notes: Audio/video streams are the encoded media and their applications. H.501 messages in H.225 sessions between gatekeepers allows the H.323 network to resolve addresses across administrative domains.

352 × 288 pixels) with ADPCM compressed voice in one direction and Quad-CIF (four times the pixels) with high-definition voice in the other.

This standard defines four logical components within the architecture. Physically these components may share a hardware platform, but are often distributed over multiple servers.

5.5.2 Gatekeeper

H.323 terminals may communicate directly with each other if they hold the necessary address information. This arrangement lacks features such as admission control, billing, and routing for traffic engineering. Larger systems contain too many addresses for a terminal to hold, making a gatekeeper (GK) necessary.

When present in a network, the gatekeeper controls all aspects of calling. It is a software process that registers end points, supplies connection information to calling and called terminals, tracks bandwidth usage per link, compiles billing information, and of course arranges connections between a calling end point and a called end point.

All of the devices controlled by a GK constitute its zone. A zone has only one GK. The interconnections among multiple gatekeepers in a network is not part of H.323 but is in Annex G of H.225.0.

Connections among zones (Figure 5.14) allow GKs to query each other about the reachability of subnetworks or individual end devices. Connectivity often is configured administratively, but H.225.0 allows for a discovery process under any of several topologies:

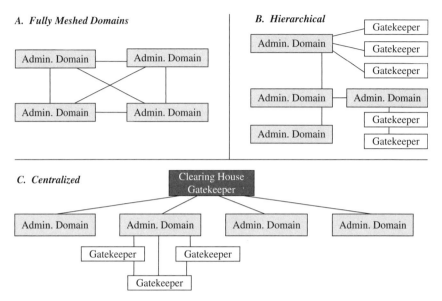

FIGURE 5.14 Networking among H.323 gatekeepers in a large network of multiple administrative domains takes any of several forms. Within a domain, multiple GK zones also network in similar topologies.

- **Star:** one gatekeeper, the clearing house, communicates with all other GK's, but they query only the clearing house.
- **Hierarchical:** the connections form a tree; a GK may respond to queries that reach it when seeking information from a zone further from the root of the tree.
- **Fully meshed:** every GK signals directly to every other in the network.

GKs operate several protocols on the IP network side:

- H.245 for control functions including learning terminal capabilities and opening media channels.
- H.225 for user/device authentication, call setup and clearing, admission control for new calls.
- H.248 for control of media gateways

For the descriptions here, H.225 packets contain only one message. Annex E of H.323 offers a multiplexing capability within an H.225 protocol data unit which would reduce the number of packets by combining sessions for transport in shared packets (Figure 5.15). However, there are 18 different header formats to accomplish that multiplexing, which seems to be more detail than needed here. The discussion contains good drawings for the header layouts, but these do not include the lower layer protocols, UDP or TCP.

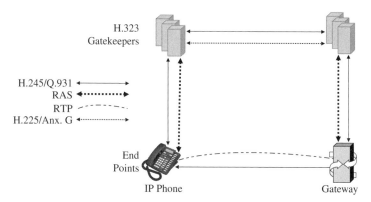

H.323
Gatekeepers

H.245/Q.931
RAS
RTP
H.225/Anx. G

End
Points

IP Phone   Gateway

FIGURE 5.15 Protocols used by H.323 gatekeepers, in general.

A terminal that downloads a configuration file or updates firmware at startup may learn the address of its gatekeeper that way. An auto discovery process is more flexible and allows a terminal to change its gatekeeper upon a failure. Terminals send a Gatekeeper Request (GRQ) message to the multicast IP address and the well-known port for RAS. A GK may respond with a rejection (GRJ) or a confirmation (GCF). Either response may contain an address for an alternate GK or an assigned GK that the terminal must use. A terminal that receives more than one confirmation chooses where to register, based on its own design, but must register with only one GK. The H.323 well-known port is 1720. A well-known port is not needed when a gatekeeper sets up a call.

After learning the address of its GK, a terminal sends a Registration Request, which describes the terminal's capabilities, preferences, addresses, and alias names. Registrations may expire after a negotiated time to live. Either the terminal or GK may unregister the other, for reasons such as maintenance, network topology change, or manual reassignment.

When there is an interface to the public circuit-switched network or a TDM PBX the gatekeeper will deal with a gateway to terminate many signaling protocols such as ISDN, Q.sig, and Q.931. Higher level protocols such as Q.932 may be passed to the end terminal. Part of handling signaling is address or digit translation, for example, between E.164 (public) addresses and abbreviated dial strings or between IP addresses and a name or a number. These are forms of directory services.

As new calls enter the network and calls tear down, the GK may track bandwidth usage on links. If a new call would overload a link, the GK may prevent call completion and deliver a busy signal to the caller. Blocking calls can prevent link congestion which would degrade all calls using the link. Tracking active calls enables the GK to control a busy lamp field at an attendant station.

Calling and caller ID information may come in the form of an alias address directly from another terminal, or be modified if passed through a gatekeeper. The accuracy and validity of direct information can't be assured. However, if signal messages pass through a GK, it can mark an ID it changes as "network provided" and may indicate that an address supplied by a terminal failed a validity test and isn't guaranteed.

5.5.3 Gateway

The media gateway (MGW) function for H.323 systems is essentially the same as for any other, such as SIP. MGWs terminate calls from different networks types and perform all the transformations in signaling and media content to allow for seamless communication. For example, caller ID information from the PSTN can be converted into a connected party display on the IP network.

The gateway may contain the capabilities of the multipoint control unit (MCU) to create conferences from connections on either or both networks (see below).

MGWs that terminate SS7 are carrier oriented and unlikely to exist in an enterprise network. Such a "trunking gateway" is invisible to the other devices in the network because it appears to whatever connects to it as a peer from the other network. On the PSTN side it appears to be a tandem switch (one that connects other switches and does not serve customers on direct local loops such as PRI's or analog lines) while supporting connections to phones and video terminals.

A gateway may contain the functionality of the GW controller or the GWC may reside in a separate device. The MGC reports its capabilities and available capacity to the gatekeeper in Setup and Release messages, among others, so the GK can route traffic, control congestion, and avoid failed equipment. The MGC function can query a MGW with an Audit message to obtain current configuration and available capacity.

Carriers who offer circuit-switched services on a packet network place access gateways on the edge of an IP backbone, facing the customer. User equipment sees only a plain old telephone service (POTS), digital local loops, or another legacy interface such as ISDN. This topology can also apply to an enterprise network that retains a PBX and relies on ISDN trunks to the PSTN. This form of deployment is likely to disappear as SIP trunks displace ISDN for carrier access and older telephone equipment retires.

5.5.4 Terminal

The H.323 terminal is a computer with audio and possibly video components as well as other features. It places and receives calls on the IP network. A terminal function is required in a gateway, where the audio components are appropriate for the other network: ISDN, T-1, E-1, or POTS interfaces.

In audio, H.323 requires every terminal to be able to send and receive PCM (G.711) according to both mu-Law and A-Law profiles. Additional codecs are signaled in H.245 messages when exchanging capabilities with other devices. These include G.723.1 for audio conferencing and G.729 for terminals on slow links (originally 56 kbit/s or less).

For data exchanges with other H.323 terminals and similar devices, the default protocol is T.120 on default port 1503. A one-way or two-way channel is optional, but when invoked is part of the H.323 call using an open-LogicalChannel message. Multiple options exist in the setup procedure to allow either end to open the channel, or for on end to ask the other to receive a connection request later.

Terminals configured with sufficient information may set up calls directly with other terminals. However, most network include one or more gate-keepers (GKs) where terminals register and obtain call control services. See above.

5.5.5 Multipoint Control Unit

Audio conferencing in an analog central office is relatively simple, requiring some volume controls on the inputs to a single electrical contact and amplifiers to take the mixed signal back on each line. In the digital world, the equivalent task requires a powerful computer, typically a digital signal processor (DSP). The DSP analyzes each stream of PCM bits coming into the MCU and may suppress channels that seem to hold no more than background noise, to min-imize the noise level in the combined audio feed. After converting all signals to a linear format, the DSP mixes the active channels by adding them together (either digitally or in analog form).

The MCU in an H.323 network performs the processing to mix audio and video inputs, then reflect the content back to participants. A good digital bridge is not easy to design and build. No terminal is supposed to receive back its own audio input—the bridged audio output to each participant may be unique. When mixing video, the MP needs to maintain synchronization with audio inputs—this means generating new time tags for the RTP packet headers.

An MCU must contain the multipoint controller (MC) function to help terminals negotiate common parameters for the call and to decide which source will feed a multicast broadcast. The mixing and processing function resides in the multipoint processor (MP), most likely part of the same MCU. The MC decides on how many streams to broadcast simultaneously. If the MP and MC occupy different devices, the communications between them is not standardized in H.323.

Each terminal can choose to receive all or a smaller number of streams, for example, if limited by its capacity to decode many inputs, and indicate in an H.245 message to the MCU how many it wants.

To do that job, the MCU must be in the media stream. That is, a conference call of three or more phones must plan to include an MCU. The gatekeeper will do that planning if the call is created as a conference. All of these are possible.

5.5.6 Call Procedures

A terminal communicates in multiple ways, on multiple protocols:

- A terminal registers with a gatekeeper on a H.225 RAS channel, an unreliable connection (on UDP). A terminal can indicate separately the features it needs, desires, and supports.
- Control signaling (H.245) uses a reliable channel (on TCP) from each terminal to a gatekeeper if present, or to a peer terminal.
- Call management or routing for a terminal uses H.225 (on TCP).
- H.323 calls establish a T.120 data connection (using T.123 and T.125) when needed for information other than voice or video, such as sharing still images.

A call is defined by the H.225.0 signals. To understand all of H.323 call signaling, five phases help organize the functions logically:

1. **Setup:** terminals request access at a gatekeeper (if present), determine if call signal messages will pass through a gatekeeper or not, form and send Setup message (which may include Fast Start information).
2. **Initial signaling:** devices determine their master/slave relationships with each other, exchange capabilities, create H.245 tunnels if appropriate.
3. **Open media channels:** based on H.245 procedures, create conferences if appropriate.
4. **Call services:** change bandwidth of a media stream, manage status, expand or shrink a conference, or apply supplementary services described in the many H.450.n Recommendations.
5. **Clearing:** close control stream then media streams, RAS updates status with gatekeepers.

H.245, *Control Protocol for Multimedia Communication*, describes the syntax for how to use a session to declare capabilities, which applies to many protocols. Each terminal opens only one H.245 channel with another device, which stays open indefinitely. This channel allows terminals to:

- Determine which end is the master and which the slave for purposes of protocol management.
- Exchange capabilities and preferences.

- Open and close logical channels for media and set them as bi- or unidirectional.
- Request an operation mode.
- Determine round-trip latency.
- Perform maintenance such as loopback.

Terminals use H.245 procedures and statements within H.225 messages to inform the gatekeeper or other terminals about the codecs it supports, the number of similar media sessions it can support, and so forth. The goal is to ensure that terminals don't waste resources by attempting to create a session that its peer can't support.

When exchanging capabilities, a terminal organizes its feature sets hierarchically. The first layer is a list of modes, such as audio codecs, that are available but mutually exclusive. That is, a terminal may support different kinds of voice or video, but all sessions may have to use the same codec at a given time.

Another list describes simultaneous capabilities, which includes voice, video, and data. This list may also indicate the ability simultaneously to send or receive streams with different codecs, for example. The number and variety of options is beyond the present scope.

Messages for H.245 fall into four categories:

- **Request:** requires a specific action and an immediate response.
- **Response:** corresponds to the request.
- **Command:** requires an action but no response.
- **Indication:** information that requires neither action nor response.

User input, as from a keyboard or mouse action, is delivered as an indication.

Some video codecs produce a "layered" encoding in multiple parallel streams. One packet stream contains the base media or minimum resolution. Additional detail derived from the original source is coded in separate streams which are transported on different RTP sessions. A terminal may support only part of a layered encoding, declining to receive the other RTP streams. For higher quality video the terminal accepts as many streams as it can process or as many it needs to reach its maximum image resolution. On a multicast session, terminals that don't take what they can't use will save bandwidth on the network.

H.323 terminals multiplex connections, often putting more than one on a shared IP address. The terminal opens a logical channel with an H.245 message. Additional signaling may be carried inside RTP packets on that channel, or signaling may be conducted over each of the multiplexed sub-channels. The distinguisher among the sessions is a transport service access point (TSAP), which may be a UDP or TCP port number. There are "well-known" TSAP values for the RAS service on a gatekeeper and for the call-signaling channel on a terminal.

To originate a call the terminal sends an Admission Request (ARQ) message to its GK, indicating the desired bandwidth. The GK may lower the maximum allowed bandwidth it its response, the Admission Confirm (ACF) message.

In keeping with the tradition of flexibility in its Recommendations, H.323 allows about a dozen variations on what messages take which paths shown in Figure 5.15. Systems may route the various sessions (call control, capability negotiation, and media streams) through or around the GK.

In the simplest case without a gatekeeper in the network (Figure 5.16), the end points communicate exclusively with each other. The procedure is simple and requires few messages.

When a GK is present, end points are required to register with a GK and use it to get call setup information (Figure 5.17). Other sessions and streams may pass directly between end points. Here the GK functions much like a SIP proxy that finds a called party, informs the calling end point, then drops out of the call. While the GK doesn't require great computing power, it is "out of the loop," so it can't enforce policy and may not be able to gather accurate billing records (other than setup and take-down times from the RAS messages). When there are multiple GKs and zones, the GKs network among themselves. Each end point sees only one GK.

The most common approach is for all control and signaling messages to pass through the GK, but media streams follow a direct path between end points (Figure 5.18). At the high end of the control range, all packets from all streams pass through the GK, including media (Figure 5.19). The media traffic imposes a high burden and is not necessary for most function. However, when a GK is combined with a MCU or MGW, then media must pass through the combined device to obtain the other services.

The cases above range from simple to complex. H.323 contains a dozen variants where the terminals may be registered with the same GK, different GKs, or not at all. Either, both, or neither GK may be in the signaling path and or the media path.

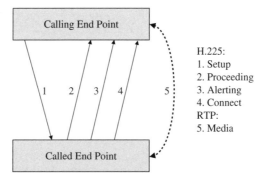

FIGURE 5.16 Simplest H.323 signaling: Setup and Connect signals passed directly between terminals.

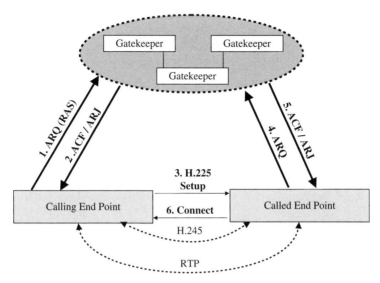

FIGURE 5.17 End points register with GK by signal and exchange media directly. If GK rejects access (ARJ) the call fails.

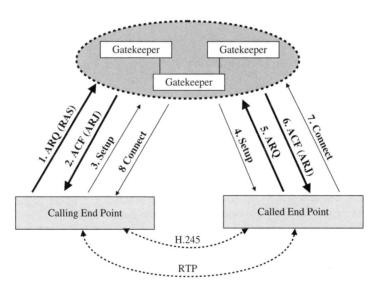

FIGURE 5.18 Call signaling routed through the gatekeeper, with media and session control on a direct path.

Terminals update their status on the RAS channel when working through gatekeepers. Thus calling end point 1 starts with an access request on the RAS channel, which is confirmed before placing the original Setup message to called terminal 2.

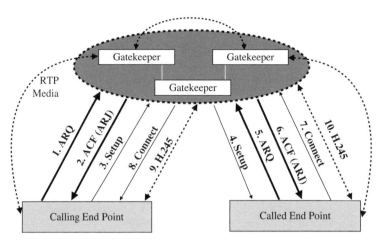

FIGURE 5.19 One of the complex scenarios for H.323 terminals registered with gatekeepers and sending signaling messages and media through the GKs for conferencing, transcoding, or other processing.

Term 2 assures GK 1 with a Call Proceeding response, then updates its status with GK 2. The Facility message tells GK 1 that signaling messages should be sent to GK 2. GK 1 then repeats the Setup message, but directed to GK 2, which passes it to term 2. If there is a delay, another Call Proceeding message may be sent. Term 2 updates its status with GK 2 (it will be busy on the line that takes this call). An Alerting message prevents the call from timing out at the GKs or term 1 until the call is answered, which triggers the Connect message.

The other configurations are not hard to extrapolate from the behavior of these examples.

To keep track of these messages for a call, particularly when a device handles many calls at once, the messages are marked in several possible ways:

- Access tokens are binary numbers that the GK dispenses to terminals at registration. A terminal includes its token in all call requests (Setup messages) to confirm its authority to use the network (being registered). End points may exchange tokens directly, to help confirm identity. A terminal may address a call using the called party's token rather than a transport address—this keeps the address private from the caller. GKs match tokens to addressees through the registration data base.

- The calling terminal generates two call reference values (CRVs) that it uses in all messages related to the same call. One CRV is for the RAS channel, the other for the H.225.0 control channel. The peer includes these CRVs in responses whether the signaling goes terminal to gatekeeper or terminal to terminal. Each pairwise connection requires a different CRV.

- The Call ID and Conference ID (CID), in contrast, are the same for all messages involved in a call. That is, messages to and from any terminal on a call and their gatekeepers include these numbers.

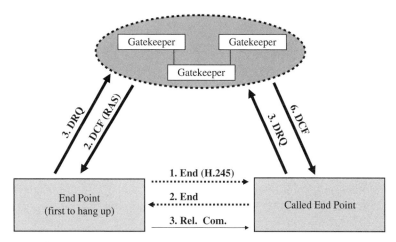

FIGURE 5.20   Call clearing between H.323 terminals with gatekeepers.

The Fast Start procedure collapses the signaling to as little as one round trip of messages. The caller sends a Setup with a FastStart element that lists all of the information needed to open the logical channels, select codecs, and establish other H.245 parameters to start media flows. The receiver may reject a fast start attempt or accept it in any message up to and including the Connect response. Either terminal may invoke H.245 procedures to configure a feature at any time during the call.

Ending a call has fewer options, but multiple ways to hang up. The three-way handshake ensures both end points have a confirmation for the procedure. End points may be registered with the same or different gatekeepers.

If there has been any H.245 activity, that channel is closed first, followed by the media sessions and the H.225.0 control channel. If there is a gatekeeper, the terminal updates its status (reports release of bandwidth, no longer busy, etc.) by sending the Disengage Request (DRQ) message to its GK, which responds with a Disengage Confirm (DCF) message (Figure 5.20). The call is then considered cleared.

Gatekeepers may clear calls by originating a DRQ message to one of the terminals, which then issues the End Session Command. The terminal that receives the DRQ responds with the DCF only after the call has cleared to the other terminal.

Many PBX functions such as call on hold, transfer, caller ID, and call waiting are described separately in a series of ITU Recommendations, H.450.n where n runs from 1 to 12 at this time. These detailed procedures will not be covered here.

RFC 4733 defines how to encode and transport signaling tones such as DTMF, MF, modem, fax, or busy in an RTP stream. A distinct payload type distinguishes tones from voice. Information about supported payload types is an important consideration for a system purchase.

In addition to specific payload types, there is a mechanism in H.225.0 to tunnel any protocol by including its payload in an H.225.0 message. A gateway can use this method to deliver signals from a foreign network, proprietary PBX signals, DPNSS, ISDN ISUP, and Q.sig to a gatekeeper or other entity that can interpret them. A particular case is when an H.323 network transports T.38 packets of facsimile information.

5.6 DIRECTORY SERVICES

LDAP, the Light-weight Directory Access Protocol (RFC 3377), offers an interface to a database that is widely supported by clients and servers. An H.323 gatekeeper could use the same LDAP server. Other data base access methods are used for VoIP, including SQL and Active Directory (from Microsoft).

Who populates the database varies across firms. Small organizations might use automatic UA or terminal registration, letting the end user insert additional information. Larger organizations will take data from Human Resources, either directly or by transferring records to the UC directory. Where security is a prime concern, the ability to remove users promptly from the phone directory will tend to tie the UC directory service closely to the HR source.

End devices such as IP phones, user agents, terminals, and conferencing appliances find the IP addresses of users from the internal directory. The client software or IP phone may aid the search by listing frequently called people, collecting recent inbound and outbound calls, or offering access to a database via queries that can search by name, title, skill, or other attribute.

5.6.1 Domain Name Service (DNS)

A fully qualified domain name (FQDN) such as YourDomain.TLD has two parts:

- Top Level Domain (`.com`, `.net`, `.org`, etc.) can contain millions of servers and IP addresses.
- YourDomain is a name associated with a specific company or organization. Different services may be hosted in different location and have different IP addresses.

The combination of the two parts, a uniform resource locater (URL), identifies uniquely at least one server, perhaps thousands, that belong to the domain. The domain name service is a hierarchical organization of directory servers that accepts queries containing a URL and returns either the IP address for a specific service on that domain or information about where to ask next.

For example, the DNS first resolves the TLD, finding the database for the `.com` domains, for example. A second query to the `.com` DNS server returns the name of an "authoritative DNS server" for `YourDomain.com`. This server is where the administrator for the YourDomain configures the master records for servers in this domain. A DNS zone is covered by one authoritative server. There may be other DNS servers with the same information, caches or slaves, closer to the source of the query. These secondary servers may answer the query before it gets to the authoritative server. Entries in caches age out and are deleted. An entry at a cache or slave server must be obtained from the authoritative server when the first query arrives (or after an entry ages out).

A DNS server that asks other servers for information not kept locally is called a recursive DNS server. Authoritative servers need not be recursive if they answer queries only for their own domains and have no need to search for other domains. A single server may act in both ways.

DNS servers may contain almost three dozen different Resource Records (RRs), any or all of which will be found in a response to a query. The main record types are:

- **A**   the IPv4 address.
- **AAAA**   the IPv6 address.
- **MX**   mail exchange, the email server's address.
- **NS**   name servers, who can answer queries about a DNS zone.
- **CERT**   certificate, the PKI certificate for a server.
- **CNAME**   canonical name, the "official name" which the DNS client should look for.
- **PTR**   pointer record, contains a canonical name or other identifier, for further query.
- **RRSIG**   a DNSSEC signature to authenticate the DNS listing.
- **SRV**   a service locator for any of several services including SIP and LDAP.
- **SOA**   start of authority record, information about the zone and its administrator.

The MX and SRV records may return separate IP addresses for email and a SIP proxy.

5.6.2 ENUM

Users outside the organization seldom put their IP addresses into its database. For reaching external destinations, there are other ways to find contact information

ENUM, eNumbers, or Electronic NUmbers Mapping draws on the Domain Name Service to find a current IP address, for a public SIP address, for example. The SIP address is of the form `SIP:name.of.user@domain.tld` or similar. A DNS query for the SRV record will return the SIP server address, if one exists. It may be possible to locate the sip server with a generic URL like `sip.YourDomain.TLD` if registered.

The inquiry may return a TEL record as well, indicating a telephone (E.164) number on the PSTN that will reach the person whose SIP address was researched. For a fully qualified domain (FQD) with a user name, DNS may return any number of records for SIP, fax, telephone, IM, email, and other means of communicating with that party.

To reach a specific company or person, starting with a telephone number is what we've done for over 100 years. On the Internet, ENUM continues to validate that approach. As defined in E.164, telephone numbers are globally unique and recognized everywhere. International, national, and regional authorities tightly control their distribution to subscribers on the PSTN. In the United States, only registered local exchange carriers, including cellular carriers, can assign an E.164 address.

The ITU recommends that E.164 numbers be written in a specific way, starting with a plus sign when the country code is present:

+ Country Code.Area Code.Exchange.Line

+ 1.202.555.1234 or + 1-202-555-1234.

As in the decimal system, the most significant digit (MSD) is at the left in a phone number. However, URLs (and DNS) place the most significant element (the TLD) to the right. Sub-domain names are added to the left.

To make the phone number above searchable in DNS, the ITU format is processed this way:

- All characters other than numbers are removed: 12025551234
- The order is reversed to put the MSD at the right: 43215552021
- Dots between digits make each a subdomain and DNS zone: 4.3.2.1.2.0.2.5.5.5.2.0.2.1
- A domain and TLD are added, making a FQDN: 4.3.2.1.2.0.2.5.5.5.2.0.2.1.e164.arpa

The TLD of arpa, devoted to Internet infrastructure, derives from Advanced Research Projects Agency, the creator of the Internet.

A DNS query of that FQDN for its Naming Authority Pointer Record (NAPTR) will return the IP addresses, or hints on where to get them, for the services provided from that E.164 address. A typical response includes an email server, a web server, a telephone number (not necessarily the same one), and a SIP proxy server reachable from the Internet. There is no limit on the number of services and IP addresses in the DNS record.

There are issues with ENUM that remain to be worked out:

- Should all telephone numbers be listed?
- Who can ensure that entries are accurate and made by the owner, agent, or assignee?
- What will ENUM do to DNS response time, cost, size, and so forth?
- How will users pay for ENUM? (Not *if*, for sure.)
- Will e164.arpa be the long-term domain for ENUM?

6

VoIP AND UNIFIED COMMUNICATIONS DEFINE THE FUTURE

Knowing some history of telephony and understanding the new technologies should prepare you to face your specific business problems. The intent is to help evaluate, select, and deploy future communications systems for at least the next decade.

6.1 VOICE AS BEFORE, WITH ADDITIONS

The goal of UC is to improve communications, which means the mark of success will be increased usage per person. What new functions will a migration add? What effect will it have on network traffic? The direction is up, but where will you start?

Before that increase hits, you should calculate if the IP network is able to absorb voice and UC in addition to data. A PBX will provide some demand information; phone bills are another source.

When planning basic capacity, compare what you know to the assumptions that vendors make when recommending the size and number of servers. Some vendors openly state their basis for a calculation, for example, in terms of the

VoIP and Unified Communications: Internet Telephony and the Future Voice Network, First Edition. William A. Flanagan.
© 2012 John Wiley & Sons, Inc. Published 2012 by John Wiley & Sons, Inc.

number of emails, phone calls, and IM messages each person makes in a day or hour. If you don't validate the assumptions against your own data, any prediction will be a wild guess and probably wrong.

The same applies to assumptions about costs when calculating a potential return on investment. More on that below.

RFC 5359, *Session Initiation Protocol Service Examples*, collects best practices for legacy services using SIP "methods" and protocols. Watch for similar documents to emerge in the future, not only from the IETF but also from vendors, associations, and industry forums.

6.2   LEGACY SERVICES TO KEEP AND IMPROVE WITH VoIP

You can still get directory service, but 411 isn't free any more. Cellular carriers will dial the number they look up for you, either for no additional charge or for a fee. The correct time is available in most areas for a local call (but is being discontinued by major carriers). The author's audible time reading was within 2 seconds of the computer display controlled by the Network Time Protocol (NTP). These you might want to keep.

Astrology readings, gambling hints, and recorded financial advice? Might not need to preserve them. With digital controls, it is possible to block connections in the same way that parental controls can block websites. SIP phone numbers and web addresses are all URLs or URIs.

The point here is that even analog phones provide more than a voice connection. Unified Communications will provide far more functionality, some new and some carried over. You will have to decide which services your system provides, with a little help from your friends in local, state, and federal government.

Once completely uncontrolled, VoIP has become too important to leave to end users and peer-to-peer desktop applications. The call volume carried on IP networks exceeded the volume on TDM networks for international calling around 2010 and continues to increase. VoIP service providers such as Skype and Vonage avoided regulation as telephone carriers for years but became too large to ignore by taxing authorities. In 2011 the requirements for E911 location reporting and certain taxes applied to almost everyone providing voice service. Certainly enterprise users need to comply with E911 laws, which are becoming more demanding in more states each year.

On the positive side, a VoIP provider classified as a local exchange carrier (LEC) gets to control public (E.164) telephone numbers, direct inward dial (DID) numbers, and can assign them to its customers. A LEC also is entitled to handle the phone number of a customer who chooses to take that number from another carrier—though this ability to port local numbers is advancing more slowly.

6.2.1 Flexible Call Routing and 800 Numbers

What exactly does a phone number stand for? How does the network find that phone? The introduction of packet switching for voice not only unifies communications but also changes the answers to those questions.

Legacy PSTN switches present a separate hardware interface for each analog phone or trunk connected to the switch. Some of these ports are in the central office, some at remote terminals. Originally the identification of that physical port, or connector, was the directory number (DN), the telephone number in the phone book that you dialed to reach that phone. This meant that a phone number corresponded to a specific local loop that extended to a specific location from a specific switch. The phone number represented the region (area code), the central office (the exchange), and the phone line (the last four digits).

To encourage people to adopt the telephone, the Bell System business model imposed billing on the caller—there was no incremental charge to receive calls (beyond the monthly connection fee). If a business wanted to save its customers toll charges, to encourage them to call from a distance, it would obtain a local number in another area served by a different switch. A business had to lease a private line from the switch that controlled the desired number the business wanted to make available to its customers. That foreign exchange (FX) line passed through the business's own telephone central office to a local loop. FX let a business have local numbers outside the service area of it own CO. The monthly FX line charge was by the mile, so it was expensive and economically practical only over a small region or metro area.

From early on in the Bell System it was possible to call and "reverse the charges" (have the called party pay), but that required the intervention of a live operator and acceptance by a person who answered the call. The process was a bother for the caller and also very expensive for the callee. But business wanted to encourage customers to call and were willing to pay. To fit the billing system, businesses that paid for incoming calls received virtual phone numbers, the 800 numbers, which were handled automatically as reversed charges. 800 numbers changed what the telephone number could represent. Now it could be any DN, anywhere in the country.

To map an 800 number to a DN that the switches understand requires a lookup in a database (Figure 3.1). This was not difficult when only 800 numbers required a translation. A software change enabled the switch to find a DN when it received an 800 call, then use the DN to make the connection.

VoIP numbers and SIP addresses likewise represent any location. The improvement is that the new addressing reaches anywhere in the world.

6.2.2 Call on Hold

Typically applied to voice sessions, the "hold" function cuts off the audio transmission while keeping the connection. The original hold button activated a mechanical switch to isolate the hand set. In VoIP, the initiator of the hold changes the state of the media session while keeping the control session open.

In SIP, a re-INV message from either end carries an SDP body in which one of the attributes (directionality) for the session is changed to:

- a = inactive if the initiator does not provide music on hold.
- a = sendonly if the end placing the hold will provide MOH.

6.2.3 Call Transfer

A UAS that accepts a call (UA = B) can transfer the caller (UA = A) to another number (UA = C) by using INV and re-INV messages. The first session/dialog (A−B, based on SD-1 and SD-2) is put on hold (see above). Then:

- UA-B sends an INV with no session description to the transfer target (C) to create a second dialog.
- The target C responds with an OK 200 message containing an SDP offer (SD-3), which contains its contact information and capabilities.
- UA-B puts SD-3 into a re-INV message on the first dialog to UA-A, who now has C's information.
- UA-A sends an OK on the first dialog, with a session description SD-4.
- UA-B puts SD-4 into the ACK to C on dialog 2.
- UA-B ACK's A.

If the user at UA-B hangs up before C answers, the system should complete a blind transfer. If B talks to C and then hangs up, it is a supervised or attended transfer, just like the old days.

6.2.4 Call Forwarding

A UA proxy may handle call forwarding for a phone registered with it. The server may apply a filter, forwarding only certain calling numbers or calls at certain times of day, to voice mail or another destination. The end user may also tell the proxy to forward calls to a temporary location by changing the primary registration.

To effect a forward, the proxy issues an INV message to the new target phone (or terminal or server; fax calls can be forwarded too, e.g., on detecting a fax modem tone), creating a second dialog. The message includes the SDP information from the caller and the caller's contact information. It is not required for responses to come back through the proxy if it is stateless.

6.2.5 Audio Conferencing

Conference call service used to be highly profitable. The audio bridge was large and expensive, not something most enterprises wanted to own. Governments lacked capital budgets, but they could pay by the minute, per participant line, to discuss projects. Travel savings justified the cost of conferencing.

More powerful computer chips and cheaper memory brought down the cost of a bridge. Even early digital PBXs offered bridging for a handful (or two) of participants as part of the feature set.

Digital signal processors (DSPs) brought the cost down further, enabling bridges to conference thousands of lines on a call. These large devices allowed rural local exchange carriers (LECs) to offer free conferencing service. Participants called, mostly from a distance. The LEC collected call termination fees paid by interexchange carriers (IXCs, long-distance network providers). Conference participants pay for the call to the free bridge. For a fee, typically paid by the host company, the bridge owner will let participants call on a toll-free number.

In the improved UC arena, conferencing has become standard at no additional cost in both hardware and software products. It may reside in a media gateway, drawing on DSPs in the hardware. Or a software bridge may run on a call control server or on a dedicated device such as the H.323 Multipoint Control Unit (MCU).

Distinguishing differences among UC conference systems that can affect your business are:

- Number of ad hoc conferences at one time.
- Ability to schedule conferences to reserve resources for large calls.
- Total number of lines simultaneously in one or more conferences.
- Controls available to a conference moderator:
  1. Mute all lines or individuals.
  2. Choose video feed to distribute; from participants, speaker, or designated source.
- Authorization or authentication methods to control admission to a conference for privacy
- Facility to distribute documents or other files, via on-demand download or pushed by the host

6.2.6 Video Conferencing

Video conferencing scored high among end users when asked about the features they most desired in a Unified Communications environment. Video to date largely has been either one-way broadcast to a large audience (announcements, coverage of live events, speeches, etc.) or a meshed exchange of images among a relatively small number of users or sites. How you want to use video leads to another item on the previous list:

- Video procedures and image switching/combining during a conference or broadcast.

Three- to five-way video conferences allow all participants to view each other with at least a quarter of the screen per site in a static configuration.

Among larger groups, the best results come from switching the main feed to whichever site is producing the audio at the moment. The more sophisticated video bridges or Multipoint Control Units (MCUs) can tie the video feed to the current audio source.

Telepresence goes further by applying stereo sound imaging to place a speaker's voice with the image.

6.2.7   Local Number Portability

A later and much larger version of the 800 number data base now includes more area codes for toll-free numbers (866, 877, and additional free-call exchanges as needed). The same technology potentially applies to every phone number because of the US national policy called Local Number Portability (LNP). The Federal Communications Commission requires most carriers to support LNP by giving up or receiving the number of a customer who wants to change service providers but keep the phone number.

It's called local portability because the original scheme restricted the change to land lines in the same area code. Later cellular numbers were added. However, with the prevalence of unlimited long-distance calling, the importance of having local numbers or 800 numbers has declined.

Nevertheless, if you've grown attached to your phone number, or spent heavily to advertise it, you probably want to keep that number when moving from TDM service to VoIP. The FCC says you can do that. Routing calls from the PSTN to a VoIP system uses the same database that controls 800 calls. Either the called or calling party's carrier may operate a gateway that receives these calls and places them on an IP network. You may retain a gateway that receives calls on traditional TDM trunks, until replaced by SIP trunks.

6.2.8   Direct Inward Dialing, Dialed Number Indication

In the PSTN, dialed number indication (DNI) allows a call through a PBX (private branch exchange, the enterprise phone switch) to ring directly on an extension without requiring action by an attendant. Telcos call this feature direct inward dialing (DID) or dialed number indication service (DNIS). DNI lets the PBX route a call to a specific phone inside an organization.

DID depends on assigning a public (E.164) directory number (DN) to a phone. LECs charge a small monthly fee for DID numbers, which are assigned in blocks that may not be related to the primary phone number. Phones on a PBX that lack a DID number can call out by requesting a trunk (dial 9), but incoming calls must pass through the attendant. It is up to the customer (you) to have DID numbers for all extensions in addition to a general number that usually rings at an attendant station.

ISDN digital trunks from the CO always carry DNI in the call request in the format of the Q.931 packet signaling protocol. A PBX with a PRI (an ISDN trunk) interface understands the message. A media gateway that receives a

PSTN call extracts all the information from the Q.931 message and passes it in a SIP message to the call control server or media gateway controller (MGC).

If the central office switch sets up a DID call to a PBX on an analog trunk, the CO switch treats the PBX like another PSTN switch on a tandem trunk. The CO passes DNI to the PBX (or the MGW) during the 4 seconds after the first full ring (which is 2 seconds of 20 Hz a.c. imposed on the line to announce a call). COs use one of several signaling methods to send DNI on POTS lines:

- DTMF digits, TouchTone sounds.
- An asynchronous data protocol message of ASCII characters at 1200 bits/s. Frequency-shift keying, a V.23 modem signal, sends tones of 1200 Hz (1 or mark) and 2200 Hz (0 or space) to represent binary digits. There are detailed signal and timing requirements. The message includes the date, time, and a checksum to confirm accurate delivery.
- Multifrequency tone pulses, where each pair of six audible frequencies (odd 100s from 900 to 2100 Hz) represents a dialed digit (or a function, e.g., coin return). This older mode is as common as pay phones—not very.

Media gateways may be configured to receive DNI information from the PSTN and will pass the called number to the UA server that routes connections. The DNIS method on the MGW much match what the central office switch uses. In addition the call control signaling must also match. MGWs, in general, support FXO, FXS, loop or ground start, and possibly wink start and reverse battery. The carrier will supply this information about legacy trunks.

Reviewing data sheets for MGWs, it appears that not all will accept in-band DNI information (DTMF, MF). Those with only digital interfaces (T-1, E-1, ISDN) seem to prefer ISDN D-Channel signaling over channel-associated signaling (robbed-bit signaling) to receive DNI.

SIP trunking takes a view of DID and DNI similar to ISDN. All SIP calls are DID, targeted to a specific recipient. A SIP address, a URI of the form SIP: name@domain.tld or similar, identifies the called party so the recipient UAS always has a specific identity to locate (or group of phones). The attendant is just another specific DID address.

6.2.9 Call/Message Waiting

Media gateways in front of a PBX can recognize the signals for a waiting message:

- 100 V applied to the loop.
- Stutter dial tone.
- Proprietary signals to digital phones.

A SIP message to the IP phone turns on the light or triggers the display. For phones without a display the system can create a "stutter" dial tone.

Call-waiting tones and messages from a PBX are converted to a SIP Notify message. The recipient can see the information on the phone' display if it has one. If the target phone is on an FXS port, the gateway generates the message waiting indication (MWI), an audible tone heard only by the one party.

6.2.10 Call Recording

One reason to focus voice traffic at one point in the network is to record conversations—for "quality control or training purposes" as well as documenting financial transactions such as brokerage orders.

Several vendors offer customer premises equipment (CPE) for recording. Having your own system allows you to tag calls with date, time, employee number, customer phone number, internal extension number, and other data to facilitate retrieval on demand. Carriers and VoIP service providers who host call control may also offer call recording.

A choice between a service bureau and you own equipment will be influenced by cost but also by how often recordings need to be retrieved and the relative ease and speed to recover a specific conversation. Concerns about confidentiality may point to on-premises recording.

Many routers, media gateways, and session border controllers have the ability to duplicate media streams and send the copy to a recorder. There the packets are saved to disk. As with most collections of information, the disks will fill. Anticipate running out of disk storage with a policy to discard earlier recordings or to back them up to another device or medium.

An example is the Alcatel-Lucent RECORD suite of software. It will record not only the voice element but screen images of agents. It works with legacy phones and, through a MGW, IP phones.

6.2.11 Emergency Calling (E911)

When 911 appeared, the phone company programmed a translation into the CO switch for your phone. Dialing 911 sent the call to a local public service answering point (PSAP). That is, 911 was translated into a DN for a line to the PSAP, similar to the handling of calls to 800 numbers. PSAP lines are configured for centralized automatic message accounting (CAMA) to deliver the calling phone number, the calling line identification (CLID). An ISDN PRI trunk also will do the job in a different format. The idea was that the PSAP could call back if necessary.

At first the choice of PSAP reflected the physical location of the telephone line demark for the CLID. That is, the address where the telco terminated its local loop on inside wiring was reported to the PSAP. This is all the detail that a LEC is certain to know in every case. For a single-family home, the location is clear. A reverse directory lookup gives the address.

In a multitenant building, commercial or residential, the demark could be inside any office or apartment at the address. It's still reasonably specific because the line carries a customer account name or apartment number. But things got messy.

The CLID itself no longer indicates an address without ambiguity when:

- Campus environments terminate phone lines in a PBX room, but the extensions cover many separate buildings, often far apart.
- Landlords operate phone services in large buildings, hiding customer names and exact locations behind a shared demark.
- Large corporations occupy dozens of floors in a skyscraper, all with the same address and general phone number.

People died because ambulances went to the main building in response to a call from another location that passed through the one PBX serving both locations.

As PSAPs evolved, the CLID became less certain to be specific. For example, an extension that lacks a DID number shows the general line as the CLID. A call to that rings at the attendant station, possibly in a different building. The need was to show a CLID that represented the location of the phone rather than the demark. PBX vendors modified their switches to provide a more flexible CLID, but much of the older equipment was not highly capable for this.

The CLID became an index to Public Service–Automatic Location Information (PS-ALI). This is a database that holds actual location information on a location information server (LIS). The local exchange carrier (LEC) or its contractor maintains the LIS for each region.

For each call received, the PSAP uses the CLID to retrieve the caller's address. The screen pop requires a dip into the ALI database. At a minimum, the ALI contains a street address. A perceived need for more specific location information grew into legislation in several states, and federal initiatives for Enhanced 911. E911 makes space for at least an additional 20 characters that can further identify the location. The enhanced database record can be more specific by adding a building number, floor quadrant, office number, or a combination of pointers to deliver a more specific emergency response location (ERL).

An ERL should represent no more than a zone on a specific floor in a building. In the most demanding jurisdictions a call to 911 must generate an ERL to identify the location of the calling phone within 100 feet. Other jurisdictions' requirements can be less strict. Almost all state laws have several points in common:

- 911 calls must not be blocked, either deliberately (to prevent prank calls) or accidentally (by not planning for them in a new phone system).
- ANI (caller ID, CLID) must be sent with a 911 call.
- Specific location data complying with local law is required; more is better, send as much as possible, including the calling extension number.

The oldest PSAPs remain tied to the ALI and CAMA trunks. The large number of jurisdictions with public service answering points (PSAPs) and a push to update them to handle more modes of communications means that the state of 911 communications is in flux. Within a decade expect development of "Emergency Context Resolution with Internet Technology" (ECRIT) to change the PSAP's connections. Work proceeds in the IETF, with several drafts undergoing work. Eventually, the Emergency Services IP Network (NSInet) in each region will tie together the PSAP(s), first responders, medical facilities, and emergency offices.

Large organizations that have a security staff will want notification of 911 calls in progress sent to the watch desk, with the location. Security staff can speed medical responses by alerting check-in points to admit ambulance crews, calling and holding elevators, and so forth.

Next Generation E911 will enable a PSAP to accept text messages, SMS messages, GPS location information from a caller's cell phone, and VoIP calls via SIP with the location information in a newly defined field of the INVITE message. A client-server architecture will be more adaptable to future technologies.

A more flexible stand-in for the CLID is the emergency location identification number (ELIN). This continues to be the phone number that the PSAP can use to call back, and the index to the location database, but it no longer need refer to the caller's specific phone. The ELIN may be a CLID, but it can also refer to a location zone that includes many phones, one of which the PSAP can reach by DID. An updated PBX reports the caller ID on a 911 call as the ELIN, not the main number or that extension's DID number. A good ELIN might ring at the desk of an employee designated as assistant emergency coordinator for the zone.

An organization obligated to provide detailed location information for the ALI has several ways to conform. The choices are roughly the same for a legacy PBX or a VoIP system.

- Install POTS lines at each location, that is, in each location zone. Route 911 calls to them, and let the LEC maintain the ALI. Changes can be slow to make. The LEC updates records once per day typically.
- Purchase a "gateway account" from the LEC to gain access to the Public Safety–ALI database. Each organization manages the records for its own lines and numbers. The gateway device costs several thousand dollars. The charge to store records is about $1 per year each. Updates take effect quickly when delivered to the gateway.
- Hire a contractor to process orders for phone moves/adds/changes into updates to ALI records. This process requires the customer to buy an ALI gateway, on-site server software, and the support service.
- Adopt E911 as a service. Small and medium businesses (SMBs) can put everything on the contractor, "in the cloud." Larger firms may need to stand up one or more servers to track all of its phone locations and help handle 911 calls.

Dealing directly with the PS-ALI database works best for firms with no more than a few hundred phones and not much movement of those phones. Larger firms may want the help of specialized server software, such as a 911 manager, a LIS, and voice positioning center (VPC). Firms with wide geographic presence or many remote offices may want a contractor that tracks the more than 5000 PSAP areas by city, county, and so forth. Political jurisdictions often share a PSAP so the mapping of PSAP coverage requires detailed local knowledge down to the street address.

Alternatively, an enterprise may engage a contractor to build and maintain a database of phone locations and zones. The "full-service" E911 contractor may process all 911 calls for an enterprise. Commercial services that provide E911 location information with calls to a PSAP offer both a local database of locations, within a campus, for example, and an interface to the public LIS. Either may satisfy the requirements for E911 calls from a corporate office, depending on the extent of the private network and the costs to build and maintain lists in the two environments.

The architecture for one vendor involves:

- A 911 server on customer premises, the location information server (LIS), that maintains a database of phone locations both inside and outside the private network.
- An IP-based call control server to forward 911 calls to the contractor's data center.
- A dedicated IP connection between the enterprise call control server and the contractor's voice positioning center (VPC), that matches locations to the PSAPs that cover them.
- A SIP-based IP channel to carry 911 calls from the enterprise call server to the contractor.
- Notification to the enterprise's security office.

All 911 calls pass through the enterprise call server, which picks up the phone's location form its own VPC. The call becomes a SIP INVITE message (containing the calling number and its location) to the contractor's softswitch. That switch uses the location to complete the call:

- Pick the appropriate PSAP and its regional PS-ALI database.
- Create an ALI record for the current location of the phone and an ELIN.
- Insert that ALI record in the regional ALI database.
- Find the telephone routing number for the PSAP from the master list.
- Forward the 911 call to an Emergency Services Gateway (ESGW) in the PSTN using the national Transport Number (TN). The 911 transport network then delivers the call to the selected PSAP.

An E911 feature is built into the firmware of routers intended for small or branch office locations that normally rely on the IP WAN for telephone service.

The 911 calls don't go to the VoIP call controller but are routed immediately to a local POTS line so that they reach the appropriate PSAP. For a small site there's no effort on the customer's part to maintain the PS-ALI database.

If the call passes through a media gateway to a circuit switched line such as a PRI, the gateway inserts the ELIN as the calling phone number in the signaling message. Calls other than to 911 carry the phone's DID number for the standard operation of caller ID.

6.2.12   Tracking IP Phone Locations for E911

VoIP phones can plug in anywhere. A user can move a phone to a different floor, take it home, or pack it on a business trip. When the phone or user authenticates to the call server, it has the same phone number but a different IP address. When an enterprise tracks migrated phones, it will also update the public service ALI database (the LIS) so that the PSAP finds the location for a phone number. In a future IP environment, the mobile or nomadic device will discover its own location, for example, via GPS or a location service of a carrier.

One common way to resolve the location of an IP phone on the enterprise network relies on its IP address. The network operator assigns IP addresses in the enterprise's Dynamic Host Configuration Protocol (DHCP) server so that every subnet address range falls into a known and controlled range of physical cabling within a defined floor area no larger than that legally required for an ELI. The location radius varies widely by political jurisdiction.

Configuration of the DHCP server restricts IP address assignments available to each zone. Any device that plugs into the network in that zone will receive an IP address, via DHCP, from the IP range assigned to that zone. When the phone registers with the call server, an important parameter is the phone's IP address. A 911 management server receives notifications of new registrations from the call processor, including the IP address in each new registration. This interface may require some system integration, often performed by the two software vendors in a partnership that enhances the value of both products.

The 911 server examines the database of IP address assignments to identify the location zone, and tells the call server the telephone number to use as the emergency location identification number (ELIN) when that phone calls 911. The call processor handles 911 calls by associating the calling phone with the corresponding ELIN, not the phone's DID number or DN. Thus the PSAP dips into the ALI database for the ELIN, which returns the correct location of the phone's zone.

Because the PSAP needs the caller ID, the SIP Forum's SIP Connect recommendation for carrier trunking requires behavior that conflicts with privacy requests. When a SIP-PBX places an emergency call it must not withhold caller ID even if the caller asserts a right to privacy on that information by including a P-Asserted-Identity header.

TABLE 6.1   Two methods to identify an internal location of an IP device

	IP Address/vLAN	Switch Port
IP address assignment	By DHCP	By DHCP
Extent of smallest zone	Subnet or vLAN	Cable drop or switch port
How 911 server resolves location	Configured table of IP's (same as in DHCP server)	SNMP search for switch port with registered IP and MAC addresses
How 911 server selects ELIN	Table matches IP address to ELIN zone	Table matches switch port to ELIN zone

Another way to locate an IP phone depends on mapping physical switch ports that terminate LAN drops to phones. Each port or group of ports is assigned to a location zone in the 911 management server. When an IP phone registers with the call control server, it notifies the 911 server, which searches the network for that IP address using Simple Network Management Protocol (SNMP). When found, the port is identified and with it the location zone. This approach resembles the earliest assumption that a number represents a location.

Accuracy in placing IP phones into zones depends on keeping the list of assigned IP addresses or switch ports up to date. Precise recording of the as-built configuration tends to get neglected. An outside contractor that specializes in network audits may prove more reliable. Automated discovery software is ofter a help in IP address management.

A record associating each ELIN with a location zone must be added to the SP-ALI database. This process may be manual, but automation in the 911 server, through a gateway, speeds the work for large networks and increases accuracy.

When the enterprise's VoIP call processor sees a 911 call, the ELIN is inserted in the call request or INVITE message as the caller's phone number.

Nomadic users of IP softphones pose a particular problem. They will acquire IP addresses from DHCP servers in hotels and airports that may not have a strictly defined geographical area that maps cleanly to a PSAP. It is possible to include these IP phones if the user will input a location when on trips.

There's at least one application for that, which runs on the same notebook computer running the softphone and links to the 911 management server. This application intercepts the registration process to interpose an additional step. Before the call control server will accept a remote softphone, the user must enter an address and any additional location details into the 911 server. That server then validates the address in the national Master Street Address Guide (MSAG) before allowing the phone to operate.

Upon verifying the location, the 911 server creates an ALI record and stores it in a temporary ALI. It is only when that phone makes a 911 call that the temporary ALI record is inserted into the proper regional SP-ALI database where the PSAP can retrieve the location.

Note that the phones do not need to be IP based. The location database for legacy PBXs has been populated manually but automated procedures are also used. For example, the human resources record typically includes a phone number and a workstation address for delivering postal mail. Those data can fulfill the E911 requirement.

A PSAP may convert to VoIP, on the way to UC, before the NG911 network becomes available. The site would insert a special purpose MGW's function to terminate the CAMA trunks from the CO switch (Figure 6.1). The gateway converts a call from the PSTN in the legacy format to a SIP INV addressed to the PSAP's proxy.

The E911 call that originates on the PSTN follows the legacy procedure as far as routing by the serving office to a 911 tandem switch. The call appears on a trunk to the PSAP, shown in the example as a loop-start analog trunk.

The E911 tandem switch seizes the line, in this case by closing loop and drawing current from the gateway. The FXS device acknowledges the line seizure with a wink signal (250 ms of reverse battery). Seeing the wink, the tandem sends the MF "Spill," the digits with ANI. The gateway creates a SIP INVITE message to the PSAP. All the MF spill information goes in the SIP From: header in one string.

The PSAP operator may transfer the call by sending a hookflash (a SIP INFO message with a "hookflash" body). The gateway converts that to a wink toward the tandem, preparing it to receive DTMF digits for the phone where the call should be transferred, sent in a separate message. A SIP REFER message from the PSAP, addressed to the gateway's IP in the Refer-To URI, will transfer the call to another extension at the PSAP whose number is in the "userpart" of the Refer-To URI.

When the VoIP call in the PSAP ends, the IP phone sends a BYE to the gateway. It then reverses the polarity toward the tandem switch, returning to idle.

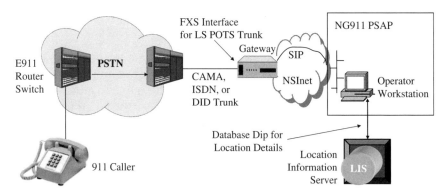

FIGURE 6.1 Gateway to receive a 911 call from the PSTN switch presents an FXS interface to appear as another PSTN switch.

6.3 FACSIMILE TRANSMISSION

Faxing remains important as a way to deliver legally recognized signed documents. Faxes are difficult to alter. By law they should be marked with the sender's name, phone number, and date stamp. A UC application can deliver faxes to individual email boxes and track in- and out-bound faxes for auditing purposes.

Unfortunately, every fax over IP implementation seems to differ in some way from every other. Hosted PBX and SIP trunking providers may handle fax calls the same as voice, using PCM encoding of the modem tones. Or they may decode the modem into a digital stream, carried in packets. Either way, they usually charge about $10/month for a fax DID number, which may be a dedicated MGW or the second port on a voice MGW.

Fortunately, the SIP Forum spent most of 2010 analyzing facsimile connections and came up with recommendations for changes to T.38. ITU is working on a draft of a revised T.38 at the same time. With the problems more clearly understood, it is highly likely that SIP interoperability testing will sort out a solution in the near future.

6.3.1 Facsimile on the PSTN

ITU's Recommendation T.30 describes real-time faxing over the public switched telephone network (PSTN) between two fax machines with analog (modem) interfaces. It's a universal format, the primary standard for almost all fax machines. The procedure is a dial-up connection originated by one machine and answered by the other. They recognize each other as fax machines by the tones they play toward each other. T-30 is digital at its core but most often travels as analog modem tones. A modem converts the digital fax format to an audible sound that passes like voice through a dial-up connection on the PSTN. Modems imitate voice in a way that PCM can reproduce.

Recall that the PSTN has practically no jitter and creates no interruptions to the audio path. Modem designs assume those conditions, which are not true for VoIP or IP traffic. Fax images, in particular, assume a dedicated voice channel for its format of a constant stream of pixel data.

T.30 persisted when communications technology converted to digital because modems were designed when channel banks had already digitized parts of the PSTN—fax machines were designed for the digital network. Almost always on the PSTN the voice channel remains uncompressed, using PCM/G.711 end to end, in part to ensure success of fax transmissions.

Carriers that compress voice can ensure success in faxing with fax detectors that recognize the modem tones and prevent compression. In a SIP environment the media gateway that connects a fax machine to an IP network should recognize the fax modem tones and re-INV to change the session from voice to T38fax. Problems exist because T.38 isn't precise on which end should invoke the change.

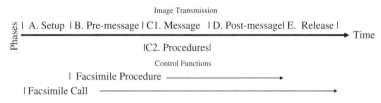

FIGURE 6.2　Phases of a fax transmission.

Fax-aware gateways respond to the difficulties with modems over IP by extracting the scan data (demodulating the modem tones) and transmitting digital information. The T.30 process assumes that the initial scanning to a digital signal follows ITU Recommendation T.4, which also includes run-length compression of blocks containing consecutive white or black pixels on a scan line.

On the WAN connection, T.30 puts the control information in HDLC frames marked by the 0x7E flag character (01111110) that is not used elsewhere. The T.4 data stream from a scan is not framed but runs continuously. Each scan line ends with a special EOL symbol (000000000001) that cannot appear anywhere else in a message. Between EOLs is a series of run-length codes, alternating between black and white, indicating a number of contiguous pixels of the same color. Here lies the sensitivity to jitter and lost packets. If the transition from black to white is confused, the received page can be illegible. On a voice channel the transmission "resets" after each line of pixels. When packetized, catching an error such as a lost EOL symbol is harder.

Transmission via modem takes place in phases (Figure 6.2) of a T.30 connection:

1. Call setup/clearing, such as dialing across the PSTN.
2. Facsimile procedures or commands by which the fax machines recognize each other and select formats.
3. Message transmission, an image encoded per T.4 or other data, sent continuously.
4. Confirmation of receipt.
5. Call termination.

Moreover fast modems have internal echo canceling, which can't train if there is even a small amount of jitter. A dynamic receive jitter buffer still allows enough jitter to pass through to spoil the connection. Network managers should turn off the dynamic adjustment feature for jitter buffers handling fax channels. Turn off VAD as well; some VAD implementations may ignore a modem tone and emit silence instead. That means end of page to the receiving fax machine, and end of transmission.

The slower modems don't cancel echo themselves, so they may be able to communicate as VoIP without resorting to T.38 gateways. However, the time to send a page increases significantly.

Greater problems for fax emerge when the voice carrier transcodes the originating PCM format to compress it into something that requires less bandwidth. ADPCM at 32 or 16 kbit/s has been used for international connections for decades. Two conversions in the path (slower, then faster) distort the reproduction of full-speed modem tones and usually make them unintelligible to the receiving fax machine. Most machines can shift to simpler modulation schemes, at slower bit rates, which may allow a transmission to succeed, but slowly.

The delay in a satellite hop is another problem, which some machines can overcome. Adding the delay of a packet network to a satellite hop make the case more difficult. Latency over 3 s (the so called T4 timer) may trigger retransmissions that overlap with delayed packets. Good spoofing of the T.30 protocol allows a fax relay system to operate with 5 s total latency.

If this is confusing, don' feel bad. Fax machines use multiple modem modulation schemes, from V.21 at 300 bit/s to V.17 to V.34bis at up to 33.6 kbit/s—on each call. The procedures and tone frequencies changed when the Recommendation was updated in 1996. Machines built to that earlier standard should be out of service by now.

Faxing over a packet network (via VoIP) offers three major scenarios under current standards:

- Real-time transmission or fax relay (T.38).
- Store and forward fax transmission (T.37).
- Revert to T.30 for the WAN portion of a connection.

6.3.2 Real-Time Fax over IP: Fax Relay or T.38

ITU Recommendation T.38 addresses the problem of how to fax in real time over a packet network by defining gateways and a protocol between them. This network replaces the audio tones in a voice channel with the digital run-length codes of the scanned pages.

Real-time transmission most closely emulates the analog operation over the PSTN. The users know they have concluded their transaction. The communications carrier never "holds" the information, so never has liability for its security under laws about health information (e.g., HIPPA) or privacy of personally identifiable information.

On the hardware, T.38 describes an interface to the packet network that carries the digital information from the scan output, defined in Recommendation T.4, in packets. A very few fax machines have a digital T.38 interface.

Almost all ordinary fax machines connect in analog (T.30 mode) on a modem interface designed for the PSTN. Rather than send that audible signal over a PSTN connection via modem, a T.38 process takes the modem tones into a device (a gateway) that demodulates the T.30 modem traffic into digital image information. The T.30 data stream then breaks into data blocks that are

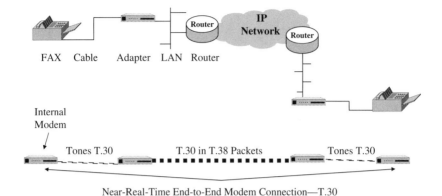

FAX   Cable   Adapter   LAN   Router

Internal
Modem

Tones T.30   T.30 in T.38 Packets   Tones T.30

Near-Real-Time End-to-End Modem Connection—T.30

FIGURE 6.3   Real-time transmission of facsimile over and IP network (T.38).

encapsulated in a special version of UDP (with error correction), then in IP packets. A digital network such as a LAN or WAN (Figure 6.3) delivers the fax stream to another fax gateway.

A modulation function in the gateway at the delivery end recreates the modem tones from the transmitted data for the receiving fax machine.

The PSTN offers minimal latency and introduces practically no jitter on an audio channel. To isolate the fax machines from the variable delays and possible packet loss of a packet network like the Internet, a fax gateway will spoof at least part of the T.30 protocol. That is, the gateway terminates or generates the calling and answering tones of the fax call setup procedure (Phase A in Figure 6.2) and accepts DTMF dial strings, emits ringing voltage, and so forth. The packetized page scan content may be protected by optional end-to-end error correction.

Lacking T.38 gateways, fax over VoIP may be treated as voice at "normal" speed (pages per minute) if using the G.711 (PCM) codec. This encoding generates 64 kbit/s plus packet headers to carry a 9.6 kbit/s modem signal—ugly, and still not guaranteed reliable. Even an all-PCM fax connection (no compressions) may fail if any part of the connection is packetized, as in VoIP, without good QoS. Nominally able to carry the modem signal, G.711 over IP often suffers enough from packet loss and jitter to make fax transmission uncertain. This is a concern because the majority of international calls in 2010 were carried as VoIP.

Codecs that compress voice channels cannot carry high-speed modem tones accurately, so the fax machines back down to simpler (and slower) modulation schemes. They too may fail, for the same reasons.

Because the fax machines see only the analog modem signals, a T.38 gateway to the LAN needs to operate in real time, or very near it. That is, the gateway creates packets at short intervals when connected to a sending machine and reproduces a constant analog signal when delivering to the receiving fax. Silence causes the receiving fax machine to hang up.

When T.38 first appeared, RTP wasn't established. A form of UDP called UDPTL (UDP Transport Layer) was standardized and widely implemented—at this time it is still the dominant protocol for T.38 implementations. Version 3 of T.38 emphasizes RTP as the better protocol.

Each UDPTL protocol data unit (PDU) contains a sequence number followed by the current or primary data block—an Internet Facsimile Protocol (IFP) message—from the fax scanner. An IFP message contains an internal sequence number. The two numbers must be the same.

If error correction is configured, there are two options:

- In redundant mode, copies of one or more earlier IFP messages follow the current IFP message in the packet. With the current and the previous interval in each packet, two consecutive IP packets must be lost to lose any scan data. With two earlier IFP messages in each packet, a data loss requires three missing packets, and so forth.
- In forward error correction (FEC), each UDPTL packet contains the primary IFP message followed by a field that indicates the number of earlier messages that are included in the FEC process, then a parity-encoded representation of those messages.

IFP messages processed for FEC must have contiguous sequence numbers, starting with the sequence number just before the primary IFP message. That is, if the primary IFP message is number S, the sequence number of the first redundant message is $S-1$ of the second, $S-2$ and so forth.

The clever FEC process creates one composite message equal in length to the longest message included in the group. The previous messages are stacked vertically, the shorter ones are padded with 0's to the length of the longest, then each bit-wide column is "parity checked": if the number of "1" bits in a column is odd, the new message has a "1" in that position; otherwise, the result is "0" for that bit position.

With this parity information from N packets of FEC fields (sent in N packets), the receiver can recreate one missing packet (one primary message) in N. There is also a more complex FEC process that generates multiple FEC messages per IP packet. This scheme protects against some error bursts that lose multiple consecutive IP packets but requires more processing power.

FEC is optional. A gateway receiving redundant IFP messages may ignore them.

To identify a fax connection, the media type in an SDP block may read:

- $m = $ audio/t38, indicating the transmission emulates fax-modem connectivity.
- $m = $ image/t38 when the data is a TIFF file containing an image of the faxed page.

Because of the wide use of UDP and UDPTL, a vendor contribution to the IETF draft of proposed revisions to T.38 offers an example of what a fax-only INV might look like between T.38 gateways (Figure 6.4).

```
INVITE sip:+1-212-555-1234@bell-tel.com SIP/2.0
Via: SIP/2.0/UDP kton.bell-tel.com
From: A. Bell <sip:+1-519-555-1234@bell-tel.com>
To: T. Watson <sip:+1-212-555-1234@bell-tel.com>
Call-ID: 3298420296@kton.bell-tel.com
CSeq: 1 INVITE
Subject: Mr. Watson, here is a fax
Content-Type: application/sdp
Content-Length: ...
v=0
o=faxgw1 2890844526 2890842807 IN IP4 128.59.19.68
e=+1-212-555-1234@bell-tel.com
t=2873397496 0
c=IN IP4 128.59.19.68
m=image 49170 udptl t38
a=T38FaxRateManagement:transferredTCF
a=T38FaxUdpEC:t38UDPFEC
m=image 49172 tcp t38
a=T38FaxRateManagement:localTCF
```

FIGURE 6.4   INVITE message for fax connection.

The first "m=" line indicates a preference for UDPTL over TCP, which appears in the second "m=" line. The response will indicate port=0 for the rejected option and a valid port number for the accepted transport. The UA software takes care of creating the message, including the Content-Length field, which was not done for this hypothetical message.

For an INV asking for RTP/UDP, the message could resemble that shown in Figure 6.5. The first "m=" line now contains RTP with the Audio Visual Profile.

After establishing an audio connection between two UACs, one of them will re-INVITE with T.38 as an attribute, moving the call to fax mode based on PCM encoding (G.711) and no silence suppression (no VAD). The SDP message in the body of the re-INV could look like Figure 6.6.

The language of Recommendation T.38 leaves some ambiguity, which allowed different implementations of fax gateways. In some circumstances they don't interoperate. For example, T.38 isn't perfectly clear on which end of a sip connection should issue a re-INV to change from voice mode to fax mode. While gateways can detect fax modem signals of various kinds, only the HDLC flag characters in the initial V.21 modem modulation will distinguish a fax machine from a data modem.

Some T.38 implementations default to a behavior where the sending MGW waits for a media packet from the receiver before starting to send page images. It's like a wink-start telephone trunk, which can be good but raises a problem when working through a firewall with NAT. The media port won't be open to that first packet from the receiver until the sender opens it with a packet from the inside. To work in this case, the sender should be configured to start sending

```
INVITE sip:+1-212-555-1234@bell-tel.com SIP/2.0
Via: SIP/2.0/UDP kton.bell-tel.com
From: A. Bell <sip:+1-519-555-1234@bell-tel.com>
To: T. Watson <sip:+1-212-555-1234@bell-tel.com>
Call-ID: 3298420296@kton.bell-tel.com
CSeq: 1 INVITE
Subject: Mr. Watson, here is a fax
Content-Type: application/sdp
Content-Length: ...
v=0
o=faxgw1 2890844526 2890842807 IN IP4 128.59.19.68
e=+1-212-555-1234@bell-tel.com
t=2873397496 0
c=IN IP4 128.59.19.68
m=image 49170 RTP/AVP 100 101
a=rtpmap:100 t38/8000
a=rtpmap:101 parityfec/8000
a=fmtp:101 49173 IN IP4 128.59.19.68
a=T38FaxRateManagement:transferredTCF
m=image 49172 tcp t38
a=T38FaxRateManagement:localTCF
```

FIGURE 6.5 Connection request for RTP/UDP in a fax connection.

```
v=0
o=faxgw1 2890844526 2890842807 IN IP4 128.59.19.68
s=FAX message
e=faxsupport@company.com
t=2873397496 0
c=IN IP4 128.59.19.68
m=application 49170 udp t38
a=t38errctl:parFEC
```

FIGURE 6.6 SDP message in the body of a re-INVITE.

blank or idle packets (also call "no-op") immediately on the port of the image session. That will open the port on the firewall for packets from the sender.

The SIP Forum has a working group on the task of improving fax over IP. A key goal is to define repeatable tests to verify interoperability and capacity to handle multiple FoIP sessions, which usually require more CPU cycles than a voice session.

Version 3 of T.38 was nearing completion in 2011. It increases the maximum speed of transmission (lowers the time per page) by standardizing on a V.34 modem, described in the ITU Recommendation of that number, which is faster (up to 33.6 kbit/s). Older machines are built on V.17 modems for page images, 14.4 kbit/s maximum. When selecting a fax machine, fax server, or media gateway with fax capability, look for the T.38 Ver. 3 capability but be sure it can fall back gracefully to V.17 modulation.

6.3.3 Store-and-Forward Fax Handling

To automate handling of faxes, the receiving process can be emulated in software on a server equipped with a fax modem. Calls from the PSTN arrive over analog or ISDN lines. The modem demodulates received fax tones into a digital form that the server converts to a file format (typically TIFF) for delivery, storage, and viewing.

If the fax server is owned by the recipient, the server is considered the end point and the arrangement is nearly real time. If a fax server is in the middle of the network—outsourced to the carrier or a third party—the message transfer may become a "store-and-forward" process described by T.37 (Figure 6.7). It is more like email for faxes. The sending fax machine transmits to a server, in either T.30 or T.38 format. The server puts the digital information on disk, where it may sit until the recipient asks for it. Or the server may deliver the message to an outbound gateway that dials a destination fax machine over the PSTN and sends the fax via the T.30 procedures.

A server holding faxes may be subject to compliance and security requirements. Like emails, stored faxes may be archived for auditing purposes and to satisfy requirements for legal discovery. The server can offer additional security by authenticating users who try to retrieve faxes, encrypting fax files, and logging when and by whom faxes are originated and retrieved.

The server may deliver the fax at a scheduled time, immediately after receiving it, or on demand. With the fax in a stored file, it is available as part of a file system or the server can forward it as an attachment to an email.

When fax servers email the image files to the user, there's a MIME type for that (AUDIO/T.38) so that the form of the attachment is understood. The recipient can then pick up the fax like any email. Other delivery options include converting the fax text to spoken words when the recipient calls in, and delivery to a specified printer.

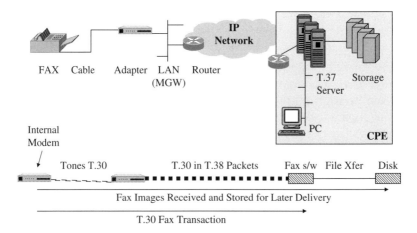

FIGURE 6.7 Fax via a store-and-forward server on a packet network (T.37).

Digital local loops such as ISDN lines will identify the called number with each inbound call. Caller ID and DID services are also offered on analog POTS lines. The fax server can associate a fax with the intended recipient through those numbers for delivery to an email address or a voice mail box.

Hosted fax services boast of large capacity to receive simultaneous faxes to the same phone number. An enterprise need not dedicate a large number of lines or machines to receiving faxes in response to a temporary rush such as a closing date, special product sale.

6.3.4 IP Faxing over the PSTN

So far, fax over IP as part of Unified Communications sounds fairly straight forward. However, the standards are not always interpreted identically by all vendors of equipment and software. There have been incompatibilities, though they are diminishing. Even the most popular brands of Fax to IP adapters don't work reliably enough for some critical applications.

The most common problems arise from the sensitivity of fax machines to delay and lost information. Too much of either and a fax machine can hang up and issue an error report.

When connected over a dialup connection on the PSTN, a voice channel has lower latency and practically no jitter compared to a packet network. IP transmission almost always increases latency (from buffering packets at multiple routers and switches across the network). If the packet network is congested, latency can vary (variation in latency is jitter).

Fax/IP systems that face regular fax machines need to look like another regular fax machine connected on the PSTN, a form of "spoofing." That is, a fax gateway needs to respond quickly and consistently to the signals from a fax machine, while dealing with a packet network that can introduce jitter and information loss.

As of this writing, the bullet-proof fax system is based on keeping the Fax over IP transmission on a LAN where the quality of service is controlled to a high level. Off-premises faxing reverts to the PSTN.

Hardware consists of fax modem cards in a server on each LAN. The POTS interfaces on the modem cards connect to local trunks. Workstations at both sites may run fax software, obtaining the benefits of message logging, easily viewable files, and flexible delivery. But each site uses the gateway to the PSTN for off-site faxing (Figure 6.8). The sending side dials up a voice connection and sends modem tones (T.30). The receiver may be another fax server or a standard fax machine.

This architecture certainly isn't elegant, or all-digital, and it doesn't rely on the Internet for inter-site transport. Many designers will reject it as inappropriate to include in a "UC" solution. However, the facsimile protocol in this case operates only over the PSTN, as it was designed. Users get all the UC benefits of routing and delivery to and sending from the desktop.

The optimal choice for you depends on how the business and the people use fax transmissions.

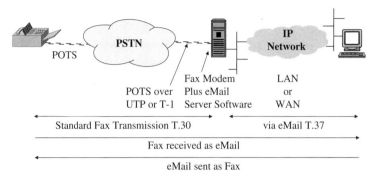

FIGURE 6.8   Sure-fire fax handling connects directly to the PSTN for off-site communications.

6.4   PHONE FEATURES ADDED WITH VoIP/UC

More than 50 projects in the IETF, additional ones in the SIP Forum, and proprietary efforts reminiscent of the PBX feature wars keep adding to the functions available on new telephone systems. Unified Communications integrates many additional features such as video, presence, and instant messaging. There are more than enough to distract the potential buyer.

The focus, then, must be on what's essential to your way of doing business. Review of sales literature will say more about emerging features. For starters, here are some things that should be available from most sources.

- **Coupling the PC and the phone:** manage the phone and calls on a PC. See who's calling, place calls, record calls, set up conferences from a web browser window or calling application.
- **Complete call lists:** not just completed calls, but missed or abandoned inbound calls—return the call with a mouse click.
- **Company and personal directories:** extensive information about fellow employees, provided from the HR database. Your contacts entered with space for as much information as you want from virtual business cards.
- **Privacy from integrated encryption:** IP phone participates in a virtual private network (VPN), IPsec, or Secure RTP—perhaps all three plus other methods—and can be configured and invoked easily from the phone or browser.
- **Choice of codecs:** for voice and video. Provides a range of trade-offs for bit rate versus voice quality.
- **Facsimile:** with a public telephone number (E.164), receive faxes in your private email box. Retrieve them from any browser or an application on a mobile phone or desktop computer. Send documents and images to any fax machine. For more details, see Section 6.3, Facsimile Transmission.

6.4.1 Presence

A survey (by Rad) revealed that end users consider presence very important in the business environment. This is not just for instant messaging (are you logged in?), but based on any information available about a user. At this writing, systems can display the busy status of the phone and meetings scheduled in calendaring servers. One vendor claimed to update presence if the keyboard is used at all.

In VoIP, status comes from the server where a user agent registers. This could be a hosted service or an "owned-and-operated" server. Many companies operate their own IM servers, for confidentiality, message archiving, or other reasons. The IM server can also track presence.

6.4.2 Forking

Find me/Follow me service in legacy switches would ring a series of telephone numbers until one answered or the call reached voice mail. Because numbers rang several times at each phone before the "ring no answer" count tripped the move to the next number, it could take significant time to run through the list.

VoIP systems improve on this feature by "forking" a call to ring multiple phones at the same time. The first to pick up gets the call. The phones that don't answer quickly return to idle status so that they can send or receive other calls— they are not tied up until the answered phone goes back to the on-hook state.

Forking proved handy in early VoIP deployments and became a favorite of people who weren't hiding behind voicemail. The caller got through whether at an office desk phone, cell phone, home phone, car phone (remember them?), or the girl friend's phone.

However, a proxy that forks an INVITE to multiple UASs will not do so unless the calling UAC provides acceptable authentication credentials for all of the UASs that ask for them. Smart proxy servers can still apply filters of various kinds to block or send to voicemail calls from specific sources, at defined times, or on designated days of the week.

6.4.3 Voicemail = eMail

Some organizations live in voicemail, some in email. UC promises to merge the two.

The technology to convert voice mail to text is speech recognition. It is highly developed in over 40 languages. When a caller leaves a voicemail, a UC system can convert it to text and deliver it via email. With cell phone displays now big enough to read the text, this can be a good way to retrieve messages in noisy areas such as airports or sports arenas.

Conversely, an email or text message converts to a voicemail via the text-to-speech feature.

The original message remains available in its first format so the recipient can check words that didn't translate or seem suspect from the context. Vendor offerings vary, so check for convenience, reliability of conversions, and costs in terms of delivery latency and required servers.

6.4.4 SMS Integration

Short Message Service (SMS) was part of the GSM cell phone technology from the start. SMS is better known as "texting." Teens live by SMS, but big business has found applications such as alerting customer to delayed air flights, low bank balance, or items on sale. An interesting possibility suggested by an SMS vendor is to update an employee's presence status, including call forwarding, through a text message to a directory server.

The background on SMS involves Signaling System 7, which has a "part" for mobile users to send text messages. The destination typically is another mobile phone. Like most telco protocols, SMS was designed to be compact—the payload is limited to 140 octets on the GSM control plane. If using 7-bit ASCII codes, the maximum number of characters rises to 180. Complex character sets (Chinese, Japanese and Korean, Cyrillic) require 16-bit Unicode so only 70 characters fit in the SMS packet. Yes, that includes the title and space characters.

Extensions and services added by carriers and entrepreneurs has grown into a huge volume of traffic. In 2010 the number of text messages exceeded the number of phone calls on cellular system.

Twitter is based on SMS messaging. Carriers offer delivery of email via SMS, where up to the first 1600 octets of the email message is split over multiple SMS packets. Sprint will deliver an SMS message to a PSTN phone via automated text-to-voice conversion. If not on a plan with unlimited text messages, the charge of $.15 each can add up quickly.

The specialized nature of SMS has been a barrier to integrating the service in business processes. As mentioned, carriers will convert and deliver a text message in other forms. Specialized service bureaus or data brokers also offer ways to connect computers (e.g., of call center agents) to SMS. Unified Communications controllers can treat SMS as just another medium. Incoming texts can be queued at a call center with email and voice calls, to answered in order.

A common interface to send and receive SMS texting is as an email. The conversion requires a gateway server, these days connected on an IP network. Formerly a modem link was part of the infrastructure, but it's no longer is needed when HTML is the transport.

Only a few of the more sophisticated call control software platforms integrate SMS. At this time applications called Twilio and Voxeo Prophecy have the capability to send and receive SMS. FreeSwitch can receive but not send SMS. Asterisk, one of the VoIP software packages that's been around the longest, doesn't handle SMS at this time.

Because SMS is essentially a service of cellular carriers, it is harder to set up private services as is possible with instant messaging.

6.4.5 Instant Messaging

IM educated users about the concept of presence. The IM application on your PC displays your contacts who are on line at the time, as you are shown to them. Almost 6% of respondents to a survey in 2011 indicated this function is important to their businesses.

To distinguish IM from SMS, the difference is in the transmission method and the kind of server that mediates the service. An enterprise may host the IM server if it wants to control its internal communications.

- IM relies on a server where users register to obtain an account, then log in to establish their presence. The server may reside at AOL, Yahoo, another third-party data center, or on premises.
- SMS is a carrier function, since it relies on the signaling system of the cellular networks. SMS "brokers" may transfer messages between the cellular side and email or another form of message transfer. This function is integrated, for example, into comprehensive call center software.

IM can attach files, and messages can be longer than on SMS. SMS doesn't necessarily report on presence for your contacts.

The protocol of IM, XMPP, *Extensible Messaging and Presence Protocol* (RFC 6120), originated as the Jabber protocol about 1999. XMPP uses streams of XML (eXtensible Markup Language) to convey information, request responses, and maintain presence. Streams may be encrypted as TLS or authenticated (using a Simple Authentication and Security Layer, SASL) mechanism within this protocol.

IM servers authenticate themselves to each other, but each server has the responsibility to authenticate its own registered users. An IM server therefore cannot vouch for the name or address of a user registered elsewhere. Address imitation and forgery are possible.

XMPP addresses can be in any properly coded character set. Beyond the simple faking possible by substituting a numeral one (1) for a lowercase letter L (l), a user can switch character sets to be more creative with "confusable characters" in imitating a trusted address or name.

Unfortunately, there seems to be no technical way to prevent deliberate confusion. An IM server can make fakes harder by limiting all names to one font and character set (see Figure 6.9).

IM relies on XML over TCP. Signing on to an IM server starts preferably with a TCP connection negotiated with Transport Layer Security (TLS) for channel encryption; other transport is possible.

On that connection the client and server open an XML stream. In a sense the procedure is the same as starting to transfer an XML document. The two ends bind their specific resources to an XML stream. However, rather than closing the connection after a file transfer, the XML "document" remains "partly transmitted" for as along as the user is logged in.

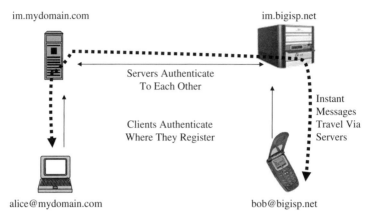

FIGURE 6.9 Instant messaging servers authenticate locally and between servers when accepting and delivering messages.

Adding an XML session fits easily with TCP as both are stream oriented. Each end of the XML connection binds a resource, like an application, to the stream. TCP and XML connections remain open indefinitely, allowing either party to push information to the other at any time.

When connected to the IP address and port of the receiver, the initiator opens a stream by sending the "initial stream header" to the receiving entity:

```
<?xml version='1.0'?>
<stream:stream
    from='juliet@im.example.com'
    to='im.example.com'
    version='1.0'
    xml:lang='en'
    xmlns='jabber:client'
    xmlns:stream='http://etherx.jabber.org/streams'>
```

The receiver replies with the "response stream header":

```
<?xml version='1.0'?>
<stream:stream
    from='im.example.com'
    id='++TR84Sm6A3hnt3Q065SnAbbk3Y='
    to='juliet@im.example.com'
    version='1.0'
    xml:lang='en'
    xmlns='jabber:client'
    xmlns:stream='http://etherx.jabber.org/streams'>
```

The entities then finish the stream negotiation process, which must include authentication via credentials using SASL. Servers may require encryption

(TLS). The order is TCP, TLS, SASL, then XMPP. For enhanced security, the devices can "forget" some of the information learned during the negotiation: the client and server may flush memory by exchanging new stream headers.

Headers may include "features/" elements, listing what is required or must be negotiated. Certain features require a restart when negotiated, for example, TLS, which is done by sending new stream headers on the same TCP connection. The XML stream gets a new ID number but is not closed.

Communications takes the form of blocks of information called XML stanzas. Any number of stanzas may be exchanged over the XML stream. There are three kinds of stanzas:

- **Message:** a push capability to send data,
```
<message to='foo'>
   <body/>
</message>
```
- **Presence:** publish-subscribe mechanism to advertise availability on the network,
```
<presence>
   <show>
   </presence>
```
- **iq (Info/Query):** request—response mechanism similar to HTTP.
  - Request:
```
<iq to='bar'
    type='get'>
   <query/>
</iq>
```
  - Response:
```
<iq from='bar'
    type='result'>
   <query/>
</iq>
```

These three elements are allowed only after completion of the stream negotiation. If either end attempts to send on earlier, the receiver must reject it and close the stream.

Users can send and receive any number of XML stanzas with any other users on the network. When a user signs off, the IM client closes the XML stream first, by sending a "stream/" tag, and then closes the TCP connection after confirming the XML closure. This procedure prevents opening a security vulnerability which arises if the TCP connection closes first.

The terminals on a long-lived TCP connections may not notice that it has failed during a period of inactivity. Therefore a stanza may be lost.

The "xmpp-client" uses port of 5222 for client-to-server connections. Port 5269 is standardized for server-to-server connections. These are the default ports registered with IANA).

6.4.6   Webinar Broadcasts

Lots of webinars in late 2010 replaced the media/analyst/consultant tours that vendors previously had to make when announcing a new product or company. The time and travel budgets saved certainly helped in the Great Recession. In many ways the web meeting is better for the recipient of information, too. For one, the recorded meeting can be replayed later to confirm details. Vendors use replays for internal training and promotional pieces.

With just two parties on a web conference, they can collaborate on writing or editing a document, create slides, or examine and discuss a complex chart or diagram in detail. It's more than a video conference.

6.4.7   Telepresence

Think video conferencing, but with feeling. The concept is to duplicate a meeting "across the table" by applying high technology and psychology. The images are hi-definition, the audio is hi-fi (7,000 Hz at least), and the lighting, camera angles, and seating are designed to contribute to the feeling of being there. Stereo sound places the speaker's voice near his image.

Broadband connectivity and H.264 video compression make the technology practical; only a half Mbit/s is needed. Dramatic drops in pricing for the components make it practicable.

Standardization proved a boon to telepresence. Different vendors' equipment interoperates. There is no need to keep the participant list to those inside one organization.

Global networks and global business leverage the savings from time and travel expense. One report claims a complete payback on a video conferencing investment in seven days of training sessions for people across the world.

A major push for the use of telepresence in the medical field shows promise of making doctors, particularly specialists, available to rural and remote areas. Thinly populated regions can't support all the specialties all the time. Doctor on demand quickly responds to a need without incurring the expense of full-time staffing at all locations.

6.4.8   More UC Features to Consider

Each UC vendor wants a unique selling proposition to convince prospects to buy its version. Look for the differentiators among features that can help your business, which could include:

- Real-time collaboration at a distance, between two users or across a group on a conference call (multicast connections). Meetings may be ad hoc, or configured administratively for a fixed team. Avaya features a drag-and-drop ability to pick participants from a directory listing to add people to a meeting quickly.

- In addition to audio conferencing on demand, UC that offers white board sharing, co-editing documents, photo distribution, file sharing, and prepared video footage as well as the speaker's image.
- Mobility whereby several vendors offer complete VoIP/UC phone features on portable devices such as netbooks and tablets as well as laptops and smartphones.
- Automated directories that combine all employees with contractors, vendors, and other contacts such as press and analysts.
- Instant Messaging integrated with system services, allowing side messaging during a conference call to any or all participants.
- Calendaring and scheduling playing an important role as at some firms where every meeting request comes through an invitation for a specific slot on the calendar. This can be too formal for some users, but it does make it clear when people are not available.
- Dashboard display of system status including usage, presence, and the modern busy lamp field.

7

HOW VoIP
AND UC IMPACT
THE NETWORK

Until UC stabilizes into a few configuration profiles, each case will likely be unique. A mix of hardware vendors and carriers will generate more permutations and combinations than any book could itemize reasonably. Setting up a gateway, session border controller, or call control server is different for each device. Good vendors provide written instructions and user manuals that consist of thousands of pages of detailed references to their command line interfaces, web portals, or SNMP MIB extensions.

That amount of detail doesn't contribute to the broad understanding that is the goal of this book. What follows is common across all situations.

7.1 SPACE, POWER, AND COOLING

Digital PBXs in smaller sizes often mount on the wall in a closet that has only nominal air conditioning from an office environment. Internal batteries provide backup power for phones, which demand little current when idle. Analog phones draw no current when on hook.

VoIP installations, particularly with many UC services, resemble larger PBXs installations or small data centers in requiring rack space for servers and greater attention to cooling. Backup power is external to the servers, and needs

VoIP and Unified Communications: Internet Telephony and the
Future Voice Network, First Edition. William A. Flanagan
© 2012 John Wiley & Sons, Inc. Published 2012 by John Wiley & Sons, Inc.

higher capacity batteries than those inside a PBX. IP phones are computers that need constant power to hold up the processor and display, in the range of 8 to 15 watts each.

Planning for a migration to UC must consider all of these physical factors as well as the configuration and operational aspects described in the following sections.

7.2 PRIORITY FOR VOICE, VIDEO, FAX PACKETS

Recall that delay degrades perceived voice quality and can break fax connections. Minimizing latency is an important consideration when migrating away from a TDM or circuit switched PBX which operates on dedicated copper cables. Putting voice traffic on a packet LAN or WAN requires careful planning to avoid objectionable queuing delays in routers and switches when voice shares the network with data.

Within the enterprise network the administrator has the authority to set priorities on different types of traffic. If a network carries both data and voice (including two-way video conferencing) the best practice is to configure routers and switches so real-time voice packets move onto transmission links in preference to non−real-time data. An exception could be made if the network is very lightly loaded, contains at most a handful of router hops between end points, doesn't carry jumbo frames (over 1500 bytes), and enjoys fast links to keep down clocking delay.

Prioritization of one form of traffic or application over all others may raise issues of office politics. Be aware that granting all packets the top priority status means that no packet has an advantage over any other. That means only best-effort service for everyone—the same as no prioritization at all.

Propagation delay is a major component of latency over wide areas. Carriers offer IP transport services designed to minimize latency, routing via the shortest or most direct physical path. But they might not prioritize some packets over others. If the carrier's backbone has enough bandwidth capacity to remain lightly loaded, the queuing delay per node may be small enough to ignore. The rapidly growing volume of IP traffic, however, points to a need to prioritize voice as a way to minimize queuing delays.

The optimal form of priority to give voice packets in most cases is absolute priority (strict priority queuing). That is, if any voice packet is queued for an outbound link, it is the next to go regardless of the size or age of the data queue. Some implementations will interrupt (abort) a long data packet to start sending a voice packet. The data packet, based on TCP, will be retransmitted automatically to ensure delivery.

Arguments against absolute priority lean toward a "fairness" queuing model, such as weighted fair queuing, where each class of priority receives a guaranteed opportunity to send a packet no matter how congested the link is. This approach is less than fair to a short voice packet, which might have to wait much longer for a maximum-size data packet. A potential threat cited by fans

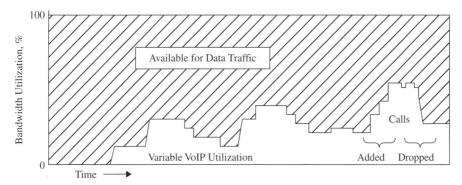

FIGURE 7.1   Voice traffic with absolute priority sees lightly loaded network; data traffic has its own queuing model for the remainder of the bandwidth.

of fairness is that any traffic with absolute priority could/will crowd out traffic of lower priority and completely starve some applications. This fear of a high-priority application appears to be based on something like a file transfer, where the bit rate can ramp up to the full link speed.

Voice traffic is self-shaping because it is constrained by the limited bit rate of a codec's output—most encoders never change speed. An adaptive bit rate codec may vary its speed at most by a factor of 2 or 4, but will never ramp up to fill a link.

The answer lies in engineering the network so that voice or highest priority packets utilize only a fraction of each link. A common rule of thumb is that voice should represent no more than a third to a half of a link's capacity (Figure 7.1).

When all voice packets have been sent first, the remaining link capacity goes to data. In effect, prioritization partitions a link into fast and slow lanes, dynamically sizing the fast lane (clear in the drawing) as needed to carry whatever voice load presents itself. A voice packet always sees a lightly loaded network with small queuing delays, minimizing jitter too.

Data always has access to the remainder of the bandwidth (crosshatched area). Within this capacity the remaining data can use a fairness, round-robin, or any other queuing model.

Note that the utilization by voice appears to rise and drop in discrete steps of bandwidth. Each call represents the output of a codec, which is constant on average over an interval as short as 100 ms during the time a caller speaks. These increments are what vendors quote as the bandwidth per conversation. In very short intervals, say 1/10 ms, the link is either 100% occupied with handling a packet or idle between packets. For capacity calculations, averages are all you need: compression reduces the average bit rate.

Calculations become slightly more complex if the codecs or the sending phone employ Voice Activity Detection (VAD) or silence suppression. These features drop packets that contain no speech information. A short packet that tells the receiver to play a background noise keeps the connection alive. The receiver generates "comfort noise" to ensure that the line never sounds dead silent. Again, the vendor can supply a statistical average bandwidth requirement for each connection. Use care when calculations involve small numbers of connections, fewer than 10, for example. Statistical averages lead to smoother results as the number of simultaneous connections on a link increases.

Links dedicated to voice, with no long data packets allowed, can operate successfully at higher utilization levels. In particular, if the voice link is full duplex so that packets in each direction flow at the same time, utilization can go quite high without encountering serious congestion. Since voice packets are all relatively short, there is no possibility of one packet delaying the next for very long.

7.3 PACKETS PER SECOND

The standard Ethernet frame has a maximum payload length of 1500 bytes. The default maximum transmission unit (MTU) on many systems is about that size.

- An email might need a dozen MTUs; 100 emails a day generates 2000 packets.
- A phone conversation accumulates 20 ms of information in a packet; that's 50 per second. So a conversation of 40 s can need as many packets as a day's email.

A router can be limited by packets per second as well as bits per second. Reading the fine print in router specifications reveals that packet may have to be significantly larger than the minimum size (64 octets) to reach the router's maximum throughput. They might have to be larger than the average voice packet.

When assessing a network in preparation for a deployment of VoIP, check routers for PPS capacity as well as bandwidth. On links that carry many simultaneous calls, consider equipment that can multiplex multiple voice or video sessions into a single IP packet. See Section 7.4 on Bandwidth for a description.

7.4 BANDWIDTH

Previous sections described methods to save bandwidth through header and voice compression, silence suppression, and protocol choice. When considering the impact of VoIP and UC on the network, those savings should be considered against the cost to implement them. The trade-offs are different in the LAN (where bandwidth is cheaper) and in the WAN (where bandwidth can be expensive).

When sizing equipment, be clear on how throughput is being measured so that you know what to expect in the boxes and on the links among them. In particular, be sure you know the difference between these two methods:

TABLE 7.1   Planning bandwidth for VoIP/UC

	Voice		Video		Virtual Desktop	
	Std.	HD	640×480	1920×1080	Min.	Max.
Bandwidth, kbit/s	170	225	540	1850	200	1000
Latency (RTT), ms	250	200	200	150	60	30
Jitter, ms maximum	40	25	20	5	20	15

Notes: Std: standard voice audio, 300–3300 Hz. HD: high definition voice, 300–7000 Hz with no compression. Values under Virtual Desktop imply a LAN-like distance between server and client.

- Leased line speeds in a traditional WAN were stated as the one-way bit rate for a two-way connection, with the same speed in each direction. That is, the rating was for one side of a full-duplex path.
- Router ratings tend to add up the inputs on all ports. A full-duplex link passing through a switch requires two ports. Adding both inputs results in a doubling of the throughput rating compared to the traditional TDM capacity measurement.

Voice conversations are full-duplex, so it is important to understand what your are calculating when engineering link and device capacities.

The ultimate question is how much bandwidth each session requires. Voice is relatively easy because it operates at a constant bit rate that depends on the codec, less any bandwidth-savings applied in the terminals and network. Video codecs may vary the output bit rate, reflecting the amount of motion in the scene, but each algorithm has a known maximum. Averages for planning the capacity of circuits are in Table 7.1, expressed in "router-rating" numbers.

7.5   SECURITY ISSUES

As part of a data network, VoIP servers may succumb to a denial of service attack, eavesdropping, and data theft. Examination of the vulnerabilities in a VoIP system confirms that it has many more potential problems than a circuit-switched PBX. To protect the system, prevent fraud, and ensure availability of phone service a VoIP solution must include a serious concern for security.

Everyone should worry about keeping important data confidential. Organizations in government, health care, financial services, and retailers who accept credit cards are all under contractual obligations, constraints, and regulations regarding privacy. Implementing a VoIP or UC system that communicates with locations outside its offices on the Internet opens an organization to many potential violations of HIPPA, Sarbanes–Oxley, credit card, and other regulations that may carry significant penalties as well as embarrassment at disclosure of customers data loss.

7.5.1 Eavesdropping and vLAN Hopping

People in the telecom industry know the government can tap a landline quite easily. The Carrier Assistance for Law Enforcement Act (CALEA) requires switch makers to build into their products a way to tap calls to or from specific numbers. The law requires carriers to allow the taps when presented with court orders (or merely a request from a police agency under antiterrorism laws). Yet few users of legacy PBX systems or the PSTN worry about a lack of privacy— as evidenced by what crooks say on the phone.

In 2005 the FCC ordered IP telephony companies to conform to CALEA, although there are technical difficulties in obeying that order. Voice packets, unless caught near the source or constrained to follow specific paths by configuration, will follow the "least cost route" established by the local routing protocol. That path may not be easy to identify and tap quickly.

However, an insider who can get close to the end point can monitor the local loop or inside wiring that carries IP and Internet access. It is not impossibly hard to eavesdrop. If voice but not signaling is encrypted, the intruder may note who calls whom, when calls involve specific phones, and many other informative details.

Snooping is also possible from LAN connections in public areas like reception and conference rooms.

From outside an intruder who gains remote control of a router can set up a diagnostic span function to monitor a circuit. The router will copy packets and send the duplicates to the hacker. Such a stream could be a conversation or any other information. Wireless (Wi-Fi) attacks of this type may be possible.

A converged network leaves little room to separate voice and data, each to its own cabling and switches. A single Ethernet drop at a workstation will serve both the phone and PC. To provide some security for the call servers, phones and PCs are often configured on separate vLANs, which use an additional header (Figure 4.3) to distinguish packets belonging to each. It's like an additional address, with 4 K options.

Modern IP phones contain a LAN switch and an extra port (or several) for connecting PCs, printers, and so forth. The phone plugs into the wall and the other devices plug into the phone. This topology positions the phone to give priority to voice and signaling packets, ahead of other traffic like PCs and printers, which is assumed to be "not voice" and thus have lower priority.

The PC plugged into a phone, on a separate vLAN, should not be able to reach the call servers, media gateways, or other UC components. However, a freely available piece of software (VLANhopper) allows the PC to fake its vLAN address. With that tool a PC can attack the voice service in may different ways.

7.5.2 Access Controls for Users and Connections

Remote VPN access poses a threat if the credentials of a mobile worker are stolen. Once logged onto the main LAN, an intruder can place outgoing calls to run up a large amount of toll fraud, look for private information, or attempt to plant malware.

Network designers typically apply two methods to keep out intruders: encryption and authentication. Both methods may apply end-to-end or on a per-hop basis. In addition, when first connecting to the network, a new device may require authentication from a proxy server or a registrar before it can insert its information in the directory or send packets to anywhere other than the registrar.

The H.232 paradigm (Figure 5.13) includes an ability of a gatekeeper (GK) to monitor the number of calls active on each link. By configuration, the GK knows the bandwidth available for voice on those links. The status of calls is kept as part of the link state information. Before completing a new call request, the controller checks the available capacity of links needed to form the connection. If a link is at rated capacity, the controller may give the caller a busy signal to prevent congestion from degrading the voice quality of all the calls on the link.

SIP signaling doesn't typically monitor the usage on each link. In the standard SIP/VoIP architecture, the call control server handles only the signaling packets, not the voice information packets, which are routed between end points by the network. To monitor and control voice traffic volume, such traffic is constrained to predesignated paths via traffic engineering or static routing. This design often works in conjunction with a session border controller (SBC) described in Section 8.4.

7.5.3 Modems

Modems? These days? Yes.

Measurements by Secure Logix, a vendor of SBCs, showed that in 2010 some of their clients had on average an hour per day of an outbound modem call on each trunk. Individuals install their own modems and dial out on voice lines to access websites, do email, transfer files, or other functions without corporate supervision or protections.

A scan of company phone lines can reveal attached modems and their types. If left connected to a PC on the company LAN, the modem's dial-in connection offers a direct path to the network, unprotected by the firewall.

A session border controller (SBC) can filter out modem calls on SIP trunks to protect voice and UC servers against this threat. Most media gateways have configuration options regarding modem traffic, from not allowing it to discouraging it by applying a policy that requires a coder, such as ADPCM, that slows modems to a crawl.

Facsimile machines use specific modems and signaling procedures that are best handled by fax relay technology, T.38, which will not in general support other modem types.

7.5.4 DNS Cache Poisoning

SIP relies on the Domain Name Service to translate a SIP telephone URI such as `SIP:joe@domain.com` into an IP address where Joe's voice service resides.

The process starts with the caller's client, which sends a query message to its local DNS server. That DNS server may be part of an enterprise network or may be provided by an Internet Service Provider (ISP). There are several firms that provide DNS service to the public. Designed as one of the first parts of the Internet, DNS originally included little for authentication—an incrementing message ID which can be guessed.

Since no low-level server could hold all the URLs and URIs of the Internet, most DNS servers are recursive servers. A recursive server holds only the most recently requested addresses or those of devices in the domain for which this server is authoritative. If a recursive DNS server can't resolve an address, it looks for it from a higher level server, the authoritative name server for the domain in question by going higher in the DNS hierarchy. Entries obtained from other servers age out and are deleted to make room for more recent requests.

Here's the problem. A recursive DNS server may send a lookup request to multiple higher level servers or to an authoritative DNS server. The protocol saves time by using the first response and ignoring the rest. If a hacker wants to redirect connections to his own server, he can send a DNS request to a recursive server, immediately followed by a fake response with false routing information. If that false response message arrives before a response from a real DNS server, the false address will be cached and the true response(s) discarded. Until the cache ages out that same incorrect address will be what resolves from all queries to that DNS server for that URL.

In short, it is possible to redirect phone calls to another location, as well as browser connections. Fortunately, this threat is not known to have materialized for voice as of this writing, but it is used regularly for Internet fraud.

To prevent cache poisoning, registrars are authenticating DNS records with a digital certificate. DNSsec is a protocol and message format that digitally signs DNS queries and responses using a public key infrastructure (PKI) that can be checked for validity. Significant numbers of domain servers started to be signed in 2010. The signed servers form a chain of trust from the authoritative name server for each domain to any caching server equipped to read the signatures.

Unfortunately, a major certificate authority (CA) was hacked and master keys were stolen. Other CAs will register names that intentionally look very similar to legitimate websites, with a substitutions of characters that are hard to distinguish (e.g., numeral 1 and lowercase l). Those fake sites are issued keys to show they are registered with an authentic CA.

Note that DNSsec does not encrypt information—that would be a separate function, probably provided by IPsec. DNSsec authenticates the source (as the legitimate DNS server), confirms the information payload has not been changed, or discards the response. In addition an authenticated response can prove that the desired URL doesn't exist, which can help avoid counterfeit sites. DNSsec gives users more confidence that they have indeed connected to the desired site, but nothing provides absolute protection.

IN SHORT: Earliest Instance of DNS Cache Poisoning

The story behind the invention of the first automatic telephone switch illustrates how cache poisoning works on DNS.

An undertaker, Almond Brown Stroger in Kansas City, MO, believed that his business was suffering because the local phone operator (there was only one) gave his calls to her brother-in-law, also an undertaker. If the caller asked for Stroger's phone number (the IP address), the operator might connect to it. If a caller asked for Stroger by name function (the URL/URI) the operator would resolve that request to her relative's phone number and send the call there. The operator (the cache) was poisoned.

7.5.5 Toll Fraud

When an outside party gains access to a VoIP system with a connection to the PSTN, calls can be placed to anywhere in the world. Criminals like to use phone service at a business because it is unlikely to be monitored, either by the company or law enforcement agencies. In a PBX the vulnerable feature is DISA, Direct Inward System Access, which allows an outside caller to appear as a local user. With the password, or on a PBX protected only by the well-known default password, you can place long-distance calls anywhere. Would you believe some people never change the default password?

Straight LD calling can run up bills of $100,000 in a weekend. Phone companies are very reluctant to forgive these charges, even if fraud against the PBX owner is proved.

Unauthorized insiders may learn LD access codes. Any person walking by a fax machine might place a call on a dedicated line that often has unlimited calling privileges. That's another reason to include fax in a UC deployment.

On a VoIP system there are ways to improve user authentication beyond a good password. That reduces this threat, but serious hackers will probably enter your network if they want to devote enough resources. Avoid this problem by not being the easiest mark in the neighborhood. When you and a companion are being chased by a bear, you don't have to outrun the bear, just your companion.

7.5.6 Pay-per-Call Scams

One of the nastiest scams is to break into a phone system, then place hundreds or thousands of calls to premium services set up explicitly to receive these calls and bill the calling company. Scam phone services charge the calling party for connecting, either a lump sum or per minute. In North America the 900 area code is devoted to pay-per-call services. They register with the telephone company so that the charges appear on the caller's phone bill.

Similar services can be set up in foreign countries that don't restrict them to special area codes. A common scam is to call a cell phone number and hang up after one ring. The missed call looks like it comes from an area code: 809, 649, 284, 876, and possibly others. While these 3-digit numbers resemble area codes, they are actually country codes for the Dominican Republic, the Turks and Caicos, the British Virgin Islands, and Jamaica. Returning a call incurs a heavy long-distance charge and a fee from the called service.

If you are a good customer of the phone company that bills you, it's possible to talk your way out of paying for some of these charges. Refusing to pay isn't supposed to affect your local phone service, and might encourage the phone company to be more careful about agreeing to bill for third parties. The FCC recommends restricting international calls anywhere you don't normally do business.

7.5.7 Vishing

Callers can spoof their caller ID to convince the target that a legitimate vendor or customer is requesting information by phone. All of the conversation goes directly into the wrong hands. This was the second most prevalent attack as of 2011, after high-volume calling.

Education is the only defense. Basic rule: don't divulge important or sensitive information on a call you don't originate.

7.5.8 SIP Scanning/SPIT

From examination of messages received, a hacker can learn details about a server. For example, different operating systems, and versions, use different headers and format content differently. Scanning techniques help hackers identify vulnerable servers (or find modems).

SIP scanning applies something similar to a VoIP system by sending an INVITE to every phone number on a domain in search of hackable devices. A high volume of calls (e.g., 5000 in half an hour) may also be SPam over Internet Telephony (SPIT) where the source is a telemarketer pushing a consumer product (they don't care who answers).

Known denial of service attacks also involve large numbers of calls, all playing the same audio track, to keep the phone system occupied. Enough calls, from a botnet, can overwhelm the access bandwidth or trunk capacity of the largest call center. Legitimate callers have trouble getting through.

VoIP systems offer a way to generate large numbers of calls easily and very cheaply. Access to the PSTN from the Internet is cheap, making a VoIP-based attack easier and less expensive than "dialing for dollars" by a room full of telemarketers. Often the goal is to present a phone number and ask for a call back.

An intrusion prevention system (IPS) should notice a DOS attack from the Internet and mitigate it. A session border controller (SBC) can identify known sources of volume calling and block those calls. Like viruses, telephone attacks have signatures. Some vendors collect them from multiple clients and share the information to improve the ability to block malicious calls. Ask your vendor how its device learns to recognize DOS and SPIT calls. As of 2011, volume calling was the most prevalent form of attack.

As a precaution, SIP end points that are not secure should not be registered under an Address Of Record that uses secure SIP (SIPS). A separate SIP AOR should be created for unsecured end points.

7.5.9 Opening the Firewall to Incoming Voice

Early deployments of VoIP sometimes opened the full range of UDP ports to allow incoming call requests. From a security viewpoint, this is totally unacceptable. The firewall must open only single-port "pinholes" for active connections.

Because most firewalls also translate IP addresses, enterprises now employ NAT traversal techniques to open ports on the firewall from the inside, on demand or to public proxy servers. See Section 8.3 on NAT traversal, STUN, TURN, and ICE.

The session border controller (SBC) is designed specifically for this situation. It can replace a firewall by handling voice and UC connections. The SBC provides at least as much security as a firewall while facilitating connections through NAT functions. See Section 8.4 on SBCs and NAT. Several SBCs appear in Section 11.3, in the chapter Examples of Hardware and Software.

A small number of user agents and firewalls participate in the Universal Plug and Play (UPnP) program. If compatible, an IP phone and firewall will recognize each other to allow the phone to instruct the firewall to allow two-way connections as needed on designated port numbers. The client opens a new firewall port for each connection or session, so may have many ports open at once. The security level for this arrangement is too low for most enterprises. It's simplicity suits residential and SOHO sites and will not be covered further here.

7.6 FIRST MIGRATION STEPS
WHILE KEEPING LEGACY EQUIPMENT

Huge investments in PBXs, telephone sets, and cable plant with less-than-current capabilities cannot be discarded lightly. The impact of a large write-down could be devastating financially. The capital budget may not allow purchase of thousands of new phones. Yet the pressure to employ SIP trunks, or adopt Unified Communications, may require some changes to the telephone system.

In many cases the migration to VoIP and UC may proceed gradually. You can replace pieces without replacing it all. The following are ways to modernize while retaining certain equipment.

7.6.1 Circuit-Switched PBX

A media gateway allows an organization to deploy SIP trunks and save on the cost of access lines while the PBX, wiring, and phones remain in place. Depending on the PBX and your future plans, you may have a choice between a stand-alone MGW and an IP interface module integrated into the PBX. If the PBX is scheduled for retirement, or there may be a use for the MGW later, the stand-alone model should win.

With adequate IP connectivity in place, the phones may be replaced gradually. If starting with SIP trunks, that too can proceed in steps. The recommended order is:

- **Outbound long distance:** no need to port phone numbers yet.
- **Inbound long distance:** port the main numbers for each site.
- **800 numbers, tie lines:** may require a larger number of ports.
- **Local calling and DID service:** entails porting or replacing all remaining numbers.

7.6.2 Digital Phones

Millions of digital phones are in place, connected to legacy PBXs. The cost of the phones, some at several hundred dollars originally, represent a capital investment that cannot be discarded easily. There is not much chance any more of getting a good price for used phones in the after market. Unless deeply depreciated, digital phones pose a barrier to VoIP and UC. During the migration from circuits to packets, those phones may continue to function with new VoIP softswitches and services via specialized media gateways (Figure 7.2).

In the longest term, expect these gateways to diminish to almost vanishing.

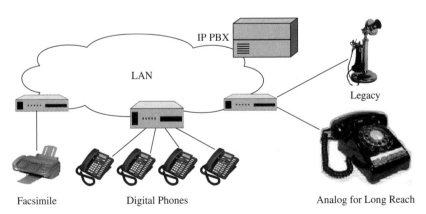

Facsimile   Digital Phones   Analog for Long Reach

FIGURE 7.2 VoIP gateway to legacy phones and fax machines.

Several manufacturers offer MGWs tailored to translate SIP signaling into the proprietary message formats of the major phone vendors. The SIP-circuit gateways behave like a standard IP phone toward the LAN.

On the downstream side these gateways provide circuit-oriented voice channels, proprietary signaling, and electrical power over existing drop cables. Ideally the gateways will displace (larger) PBX cabinets where they used to terminate the installed cabling. This migration strategy can postpone an upgrade of the IP network or LAN that might be required for VoIP.

The conversion allows the existing phone to handle most SIP functions. At this writing, there was not a SIP feature for every one of the 400 or more features available from legacy PBXs. Updates and improvements in GWs for digital phones should continue to appear for years as new features come to SIP that are the equivalent to PBX features. The amount of software development should be relatively small. You can hope for firmware updates on these gateways.

IP phones should decline in price even as the depreciated value of the installed equipment continues to shrink. In the end, the maintenance and power cost of the gateway will exceed the financial hit of changing out the phones. The digital phones, and their gateways, will then retire.

7.6.3 Analog Phones and FX Service

Where the long reach of an analog drop is needed, or where some other legacy device (voicemail, voice response system, etc.) can't be replaced immediately, a media gateway (MGW) allows that analog equipment to work in an IP environment (Figure 7.2).

For distances beyond the reach of an analog loop, early adopters of VoIP who first replaced the WAN transmission service found they needed a replacement for a "foreign exchange line." Originally the FX was a physical copper pair between central offices, allowing a phone to have a number that wasn't associated with its local CO switch. The term came to apply to a connection that allowed a phone to use a PBX from a different location.

Like an extension cord, each end of the FX line is the complement of the other. The MGW facing the PBX presents foreign exchange office (FXO) interfaces that would face a central office switch. That is, it must accept battery (the source of loop current) and ringing voltage. In the branch, the MGW shows foreign exchange subscriber (FXS) ports that generate ringing voltage and supply loop current (Figure 7.3).

If branches and headquarters had IP data connectivity with sufficient capacity, MGWs offer essentially free long distance among offices. Depending on distance and number of phones, FX over IP may be a good solution for a campus that otherwise would have to replace a cable of hundreds of pairs.

Some MGWs rely on media gateway controllers (MGCs) to act as user agents over the IP network, setting up calls using SIP. MGC functionality may reside in the same hardware as the MGW.

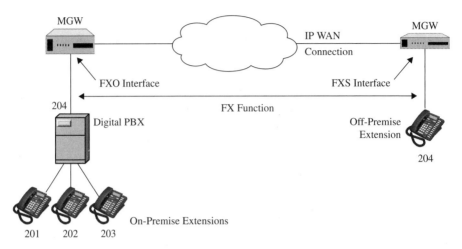

FIGURE 7.3 Two media gateways on an IP network emulate a foreign exchange line between a PBX and analog phones.

Other MGWs are smart enough to maintain a 1:1 relationship between an FXO port at the PBX and an FXS port at a remote phone. The benefit is that the remote MGW doesn't need a local MGC. The downside is that without a local MGC the MGW may not be able to fall back to a local trunk if the IP connection fails, although that is possible.

Various MGWs support different configuration options:

- **Automatic ring-down:** when a remote analog phone goes off hook, the MGWs signal the PBX by going off hook on a dedicated FXO port. The PBX then rings a predetermined extension.
- **Single dial:** the FXS gateway collects dialed digits before setting up a call by sending the dialed numbers to the FXO end.
- **Two-step dial:** When the phone goes off hook, the MGWs set up an audio connection to the PBX. When the PBX delivers dial tone, the MGWs pass this signal to the caller, who then dials. The input DTMF tones are sent to the PBX for call routing. Caution: the FX will hold the port on the PBX until the phone goes on hook

7.6.4 Facsimile Machines

Facsimile is still important to business. Firms dealing in contracts find the service necessary for transmitting legally binding signatures.

The sensitivity of fax machines to line errors, lost information, voice compression, or relatively small changes in the transmission rate of the connection (jitter) requires special attention, as described elsewhere in this book. In short, fax benefits from special gateways that address its specific needs. See Section 6.3 on Facsimile Transmission.

7.6.5 Modems

It may be a surprise, but modems are used a lot, even when Internet access is available.

Fortunately, modems were designed to operate over the PSTN at a time when the network was using digital transmission technology (channel banks) internally. Unfortunately, fax modems expect PCM exclusively, which excludes any form of voice compression anywhere on the transmission path. Such a restriction may require engineering attention. With good quality of service (QoS) for VoIP connections an IP network should handle most modem types as a standard voice call—if the media gateways encode as PCM and are part of the high-quality IP network.

8

INTERCONNECTIONS TO GLOBAL SERVICES

Many organizations venture into VoIP first as an internal service, perhaps for a subgroup of users who test a new technology like a hosted PBX service. Starting a small technology trial, the costs may not be the most important item to evaluate. However, when planning a larger deployment the costs quoted need close examination, for several reasons:

- Temporary pricing, good for one to three months, should have no influence on a decision about a long-term solution. That "saving" will disappear in a blink.
- Taxes are now applied to VoIP service, where earlier they slid by invisibly. The conversion to IP cut tax revenues paid by legacy services enough to be noticeable, so the tax man came.
- FCC authorized fees, starting with the Federal Universal Service Fee at $4.75 to $6 per line, can be half the monthly recurring charge (more than half for a cheap fax line).
- Carrier fees are creative ways to charge for overhead costs, like recovering a state gross revenue tax, right-of-way expenses, and even defending the carrier's intellectual property rights from patent infringement ($1.99/ month per line).

VoIP and Unified Communications: Internet Telephony and the Future Voice Network, First Edition. William A. Flanagan.
© 2012 John Wiley & Sons, Inc. Published 2012 by John Wiley & Sons, Inc.

As in some cellular data services, "unlimited" may have limits that when exceeded incur additional cost. Usage charges to off-net international subscribers vary by country. Note that when calling a mobile phone in some countries the calling party pays for the receiver's air time.

Even some "free" services may require that the customer specifically request them. A multisite organization that takes IP Centrex from one carrier might expect that calls between sites don't cost extra. True, but only if requested. As Verizon states on its website:

> VoIP IP Enterprise Routing ("VIPER") allows calls between Company VoIP Customer locations to be terminated without incurring per-minute U.S.-domestic or international usage charges provided both the originating and terminating locations have the VIPER feature enabled. There is no additional fee for VIPER, but Customer must order this feature to obtain its benefits.

Planning and research will pay dividends when adopting VoIP and UC. You must read all the fine print.

8.1 MEDIA GATEWAYS

The move to VoIP might be made with an IP addition to an existing PBX or by a forklift upgrade to replace the PBX entirely with a new system. Depending on the order of migration and the progress made, a location may need connectivity to the WAN and PSTN via IP or legacy access technologies. Examples are:

- SIP trunking replaces PRIs for the first step and needs to connect with the legacy PBX and phones.
- A section of VoIP added to a campus needs to work with a PBX and legacy trunks.
- TDM and POTS trunks may need to stay in place, perhaps to avoid contract penalties.

In the first case calls outside the premises requires an IP interconnection with the local exchange carrier (LEC) or an interexchange carrier (IXC) through a gateway. The other cases reverse the MGW, putting the legacy interfaces toward the PSTN or PBX and the IP interface on the LAN.

A gateway, by definition, stands between two different types of networks. The gateway participates natively in each network, making each network appear to the other as if it were the same type. The simplest way to convert between VoIP and traditional TDM channels is a local gateway, appliance, or router. This device works with VoIP packets on the LAN side, legacy signaling and TDM circuits on the other side.

Many gateway functions have existed in various forms for decades and persist in data devices that enable communications between, for example, IP

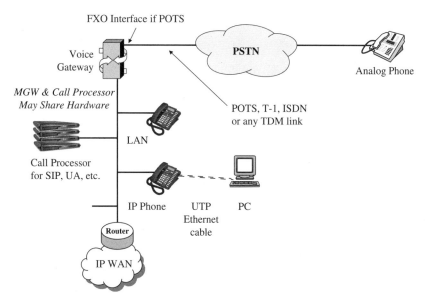

FIGURE 8.1 A media gateway participates in both networks using VoIP on the LAN and legacy TDM on the PSTN.

LANs and Frame Relay transmission links, or between devices that rely on legacy protocols, such as SDLC, and a host on the Internet.

Because LECs did not initially offer any interface other than analog local loops or digital trunks, early-adopter enterprises were responsible for the conversion gateways. Now that carriers offer VoIP directly, in the form of IP services on IP access circuits, the necessary gateways to the legacy PSTN may reside in the carrier's CO.

The enterprise MGW behaves as a host on the LAN and as a PBX to the PSTN (Figure 8.1). This gateway translates signaling messages and voice data from the packet formats on the LAN to the TDM formats and signaling required on the PSTN. The MGW may include its own controller, but the media gateway controller may be a separate device, perhaps a function of the call control processor, H.323 gatekeeper, or SIP proxy server.

The port that faces the CO switch is called, for historical reason, foreign exchange office. It is the function that faces the central office, the switch. This means FXO must accept ringing voltage (you may be shocked, literally, to learn that it can exceed 100 V a.c.; the 20 Hz frequency rattles your bones). Depending on the signaling method used on analog, the port may have to be configured for loop start, ground start, wink start, or reverse battery signaling and line supervision (Table 8.1).

Ordering the appropriate gateway and configuring it properly requires fairly detailed information about the trunks and services provided by the carrier. It is important to match your selection to the services and signaling at each site.

TABLE 8.1   Central office trunk signaling methods

Method	Description	Application
Loop start	Analog phone goes off hook and draws loop current either to answer a call or to alert the switch to a new call originating.	POTS service, phone or PBX
Ground start	Switch grounds one lead of pair in a local loop to alert peer.	Switch, PBX
E&M	E and M are two conductors for signaling, separate from the voice path; switch will alert peer by grounding or applying a voltage to its M lead, which is sensed on the peer's E lead; may carry pulse dialing.	Switch, PBX
Wink start	CO briefly interrupts loop current when ready to accept dialed digits.	CO trunks
Reverse battery	CO briefly exchanges local loop leads between positive and negative battery (d.c. power source for loop current).	CO trunks

Voice and fax gateways are found in several forms:

- **Embedded in routers:** voice modules plug into slots in the router chassis or mount on other interface modules. Works well in small sizes but becomes bulky and expensive on a larger scale.
- **Dedicated softswitch servers:** includes a call server or gateway controller as well as the MGW function.
- **Specialized media gateway:** MGW hardware platforms under a separate media gateway controller (MGC), which may be a logical function of an IP PBX. Comes in the widest range of sizes from a single voice port to thousands of digital DS-0 channels.

The optimal choice depends on the number of PSTN trunks involved, call volume, requirements for availability, and cost. There are many choices: at this writing, the site `http://www.sipcenter.com/sip.nsf/html/Gateways+Softswitches` listed more than 150 MGW vendors.

Many routers offer the option of analog and digital ports for PSTN connections. Some router hardware is optimized for voice. Many general purpose routers accept a voice module in the same place as a LAN or WAN interface card. Typically call control for branch locations resides in a data center as long as IP connectivity exists, but router now offer limited stand-alone functionality. When the IP trunk connection is down, these routers connect calls within the site normally and pass all outside call directly to local backup trunks.

Softswitches can provide similar fall back on the loss of SIP trunks, but a MGW needs a media gateway controller at all times. It's still good practice for branch offices to have some local trunk access to the PSTN, which also could satisfy E911 requirements in a small site.

A softswitch, as often understood, combines a MGW with some call control functions. It may be the only component at an office other than IP phones (and the LAN). A typical device is a PC, router, appliance, or server with voice trunk ports and a NIC. Single-site installations work well with this arrangement if high availability isn't critical or staff on site can handle both hardware and software faults.

Larger installations tend to favor stand-alone MGWs because they offer very large capacity—up to thousands of trunks. Specialized hardware is relatively compact compared to a router- or server-based solution for a large site. When built for carrier or enterprise service, a dedicated MGW incorporates redundancy of power supplies or other units to increase availability.

As a separate device, the MGW can bring huge processing power, connect to high-speed multiplexed interfaces, and service a large number of channels. It is not intelligent in the sense that it can route or control calls. The procedures between the MGW and MGC are covered in more detail in Section 5.4 on Megaco, the latest protocol for communications between the MGW and its controller.

Items to watch for in selecting a MGW include:

- Analog trunks are usually 2-wire but may use loop-start, ground-start, or wink-start procedures to time the transfer of dial digits and CLI/ANI. There are at least two forms for delivering called and caller IDs.
- Transmission level compatibility such as the length of the address header and DLCI for a frame relay virtual circuit (VC); T-1 configurations such as the form of framing (regular or extended), how excessive zeros are suppressed (B8ZS or ZBTSI), and the number of active robbed bits for signaling (AB or ABCD). There are other attributes for Ethernet access, similar to LANs.
- All IP interfaces should be ready for IP version 6 (IPv6).
- Compatibility with the carrier's method to handle telco services such as:
  1. Direct inward dialing (DID) or dialed number indication (DNI) from the central office.
  2. Caller ID on received calls.
  3. Calling line ID (CLID) sent into the trunk when placing E911 calls,
  4. Line supervision and the type of signaling on analog trunks (possibly reverse-battery wink signaling).
  5. DTMF dialed digits, which may be encoded as voice by G.711 codecs or converted to a signaling message based on the Key Press Markup Language (KPML).
- Ability to fail over to the PSTN via local trunks if IP connectivity on the WAN breaks, or if the quality of the voice service falls below a threshold.

8.2 SIP TRUNKING

Another purpose of a MGW is to make SIP trunks available to legacy systems. The orientation of the MGW for this use case (Figure 8.2) is the reverse of Figure 8.1.

When the local exchange carrier offers telephone service on the PSTN as a VoIP service, the local loop carries the same IP packets used within the LAN. Services introduced after about 2009 promote SIP signaling (rather than H.323), hence the common name of "SIP trunking" for VoIP access over IP on the local loop.

Depending on the carrier, the IP access circuit could be T-1, DSL, a radio link, or a fast channel on an optical fiber. The layer 2 protocol is determined by the access provider. For broadband access, there is a strong argument for using Ethernet as the layer 2 protocol (link level) to simplify configurations. Others will use frame relay, ATM, a tunneling protocol, or generic encapsulation in an HDLC frame.

SIP trunks are logical connections over the IP access link, using RTP/UDP/IP and a layer 2 protocol. The carrier's end of the signaling connections terminate on a proxy server, call control server, or gateway. The voice path may be the open Internet or an IP network reserved for services that demand a strict Service Level Agreement to control latency and packet loss. For connections between sites of the same organization, MPLS service engineered to provide low latency and low packet loss is a best practice. The service should be under a tight Service Level Agreement (SLA) with some serious consequences to the carrier if the QoS falls below the designated thresholds.

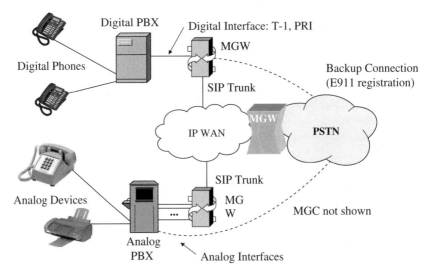

FIGURE 8.2 A voice gateway enables SIP trunking for a PBX without adopting a VoIP system. The gateway may be integrated into the PBX.

Voice connections differ only a little from data connections until they reach the carrier's backbone. There the voice packets should receive high priority. Packets enjoy high QoS either by router configuration to respond to markings in each packet for differentiated service or by routing over an engineered MPLS path.

As part of SIP trunking service, the local carrier provides gateways to other carriers. The interfaces to other VoIP services based on IP transmission most often pass through a session border controller (SBC); see below. To reach the circuit-switched public network the VoIP service provider deploys a media gateway (MGW); see above.

Before a carrier will deliver calls via a SIP trunk, the customer premises equipment (CPE) must register its signaling IP address and Address Of Record (AOR) with the service provider. The SIP Connect Technical Recommendation (TR) states that VoIP CPE must be able to initiate automatic registration with the carrier's network. This procedure is the SIP REGISTER method, described by Figure 5.5. The CPE needs to start the process to ensure that the connection can penetrate the enterprise's own firewalls and NAT functions, but also to ensure that the carrier's server can authenticate itself to the CPE using TLS. The security of that connection also guards the registration information.

If the CPE tries to register an AOR that the carrier doesn't have in its database, the CPE will receive a 404 Not Found error. If the CPE's authentication fails, the carrier may return an error message or quietly ignore the REG message. SIP Connect contains more detailed procedures for handling other response messages, but these would be handled without user intervention.

The alternative of static configuration (including bindings in firewalls and NAT boxes) eliminates the need for these registration procedures. However, the need for mutual authentication between signaling servers requires in every case that the CPE server hold a PKI certificate.

It should be clear that an IP PBX or UA proxy must support TLS. In addition to registering with the carrier, this capability can also be applied to connections to other servers in the customer's network such as directory data base servers and media gateways. DNS records for the carrier's UA should indicate support for TLS.

The SIP server as well as IP phones inside a firewall and NAT need to open a two-way channel to the SIP trunking proxy server outside the firewall. RFC 5626, *Client-Initiated Connections in SIP,* addresses this issue. A document published by the SIP Forum in March 2011, *SIP-PBX/Service Provider Interoperability* (TWG-2, "SIPconnect 1.1 Technical Recommendation"), limits its scope to URIs based on E.164 telephone numbers but offers a profile that promotes interoperability. Figure 8.3 shows the relevant network topology.

The carrier SIP server outside the firewall combines functions:

- The home or authoritative proxy that receives incoming INV messages from the public.
- The location service/registrar where the UAC registers when it comes online and joins the network.

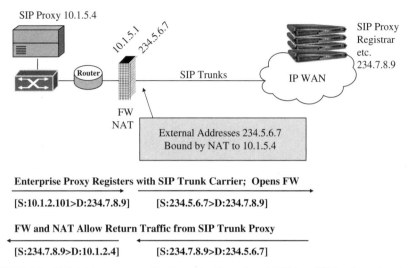

FIGURE 8.3 Maintaining a connection through a firewall and NAT to a SIP trunk service starts with a REGISTER message to open a firewall port and bind to an external address.

The message to register opens a port in the firewall and secures a binding to a public address in the NAT device. It is this public address that the registrar sees, making it available to the proxy. The UAC maintains the connection with keep-alive messages.

To minimize traffic and processing, the keep-alive is short and sent at a recommended interval of 1 minute. A TCP session should require no special attention. A UDP connection, however, may time out between keep-alive messages unless the former firewall standard of 30 seconds to time out for a UDP port opening is increased to 300 seconds. This value allows the loss of at least one UDP keep-alive packet without closing the port.

At this writing the maturity level of SIP trunking services isn't highly advanced. Confirmation testing or certification of interoperability between a carrier and specific vendor equipment has value when designing a solution. Eventually the interface should settle down to one or at most a few versions spelled out in implementation agreements such as the SIP Connect initiative of the SIP Forum and. At that time testing may not be necessary for most users.

SIP Connect 1.1 offers some guidance on what to expect from trunking services by listing features that must be supported by both the carrier and the CPE to ensure interoperability:

- Transport in RTP over UDP.
- The G.711/PCM voice codec in A- and mu-Law versions; must include type 13 in the m = audio line of the SDP block if discontinuous reception *is* acceptable for VAD (default is no, not accepted).

- G.729 codec if any compressed voice codec is offered; state explicitly if VAD (Annex B) *is not* acceptable (default is yes, accepted).
- TLS mutual authentication (based on PKI certificates) for signaling dialogs between SIP-PBX and carrier's proxy. PBX must initiate the dialog.
- If DTMF tone reception is supported, must advertise via SDP and follow procedures in RFC 4733. (Necessary because there are at least six formats for DTMF tones in packets.)
- Any end point that introduces echo must cancel it (per G.168).
- Advertise support for T.38 facsimile features if present; assumes UDPTL ability; required for any device with an RJ-11 telephone interface.

Local exchange carriers may restrict static IPs to service under a business tariff, which excludes residential Internet access, the most economical fiber-to-the-premises and DSL connections. Residential lines receive a dynamic IP from the carrier via DHCP. Business services may incur a monthly recurring charge more than three times that of a residential service for comparable bandwidth. The service level agreement (SLA) on a business service is much better than at a residence, for example, on availability and time to repair.

A static IP address is not always required to host a public VoIP server. An IP PBX that receives a dynamic IP address from a DHCP server can tell a carrier or service provider its IP address via the SIP REGISTER message. If the IP PBX is behind a NAT device, registration presents the public address to the VoIP service provider.

Various vendors call their device that connects to a SIP trunk a public server, a proxy, a mediation server, or a front end. A session border controller (SBC) or softswitch also works. They occupy the same position in the network as a media gateway, but the media (IP) are the same on both sides. For a PBX or key system hooked to SIP trunking services, the transition device is a smart media gateway that may not need a separate MGC or call control processor.

The SIP Forum's SIP Connect 1.1 defines a set of transactions, between the service provider and the enterprise signaling device, for voice only. Other media may require additional messages, formats, and procedures, some of which RFC 5626 describes. The SIP Connect document includes the concept of a "trusted SIP-PBX" in an enterprise, which may have more privileges than an untrusted SIP-PBX. For example, a trusted customer might be able to forward incoming calls to an off-premise phone or terminal.

If your carrier offers an option, automatic registration offers you more flexibility and simplifies setup. A carrier may prefer static configurations for large installation with complex SLAs or routing.

At this writing the terms and conditions of service vary from carrier to carrier. For example, a local exchange carrier offers up to six simultaneous calls on a direct fiber service, but only it you buy the package of unlimited local and long-distance calling. If you take the minimum service, paying per minute for both local and LD, you can have no more than one "line" (one call at a time) and one phone number.

A national hosted service offers IP PBX functions over SIP trunking on your existing Internet access (if it's fast enough). However, this service does not provide any phones or a media gateway, which the user must find and install. There are value-added resellers (VARs) who will do that work, for a fee.

Number portability applies, if you want to keep your existing phone identity. LECs take longer to give up a number than a VoIP provider needs to set up new service. A good migration plan first sets up SIP trunk service on a new billing telephone number (BTN). When that is working, existing phone numbers are ported from the LEC to the new service as "additional DID lines."

Dialing a number simply ported from the PSTN to a SIP trunking service creates a SIP INVITE message with the new BTN in the To: header. This means the DID information is lost. For an additional monthly recurring charge (MRC) the DID numbers can appear in the INVITE. The difference is in E911 registration: DIDs must be listed in the ALI database to appear in INVITE messages as Request-URIs coming into the enterprise. The cost uptick is about $2 per month, about the same as the per DID charge on the PSTN. There may be a separate charge to list a number in the ALI database.

If a SIP trunk terminates on customer premises in a media gateway, DID works by delivering the DID number (the last 3 to 5 digits) the same as a CO switch. That is, as DTMF tones or a data block via async modem. The DID number is delivered after the first ring on an FXS port, or after receiving a wink from an FXO port on the MGW.

8.3 OPERATING VoIP ACROSS NETWORK ADDRESS TRANSLATION

SIP signaling servers and proxies listen for connection requests on a fixed UDP port (e.g., the default 5060). Configuring firewalls to pass these packets is relatively simple—if there is no network address translation (NAT). Intrusion detection/prevention systems can examine packets on these ports to filter out anything other than a properly formed SIP packet behaving as it should.

NAT allows many hosts on private IP addresses to communicate over the Internet on one shared public IP address. The NAT "box" maps the private address (10.x.x.x, 168.192.x.x, etc.) and a port number from the transport layer protocol (UDP or TCP) into another port number on the static public IP. NAT allows many applications and users to share one external IP address on the Internet. The NAT firewall uses the destination port number of incoming messages to map the destination back to the host IP and port that originated the session and opened the pinhole (port number) in the firewall.

Firewalls assign or map addresses and filter packets in at least a dozen different combinations. Most of them differ in "address dependence." That is, the firewall may map a source address on a new outbound flow to:

- The same external socket (IP + port) for any destination.
- A unique port for each external IP.
- A unique port for every different destination socket (IP + port combination).

When filtering an inbound packet, a firewall first matches its destination address and port to the public socket of the host that opened the pinhole. Then the FW may block a packet unless its:

- Source IP and port of the incoming packet equal the destination IP and port in the outbound packet that opened the pinhole.
- Source IP of the incoming packet equals the destination IP that opened the pinhole.
- Anonymous, in which case any source socket is permitted.

Thought of as a security measure, NAT hides the internal topology of a private network from outsiders and conceals the actual IP addresses of the hosts on the LAN. This obfuscation provides some shielding from attack, but initiating connections from the Internet is harder for several reasons:

- Mapped port numbers do not correspond to standard or "well-known" values.
- Firewalls typically require an outbound packet with a specific IP + port address from inside before passing packets from outside back to that socket; external hosts can't initiate connections.
- STUN and TURN can deal with NAT on one end, but they fail if both hosts are behind NAT (see below).

Recall from Figure 5.1 that the media streams can flow between end points to minimize processing expense and latency. The goal then is to let each UA address the other directly.

When NAT changes, the IP address of a packet as it passes out from the internal to an external network the basic signaling protocols usually fail. More complications can arise when the ports are also translated (network address and port translation, NAPT).

Under the protocols that govern VoIP and UC, packets that carry voice information (the "media packets" or "bearer channels") should contain the IP addresses of the two end points (UAs or IP phones). If NAT is in the path, as is possible at one or both ends, the IP addresses used on the network between the NAT devices will be different from the addresses on the IP phones (UA hosts). The phones no longer can give each other their native addresses during call setup and expect a connection to pass voice packets. In many cases those native addresses will be private, or unroutable on the Internet (10.x.x.x or 192.168.x.x). Figure 8.4 illustrates the topology.

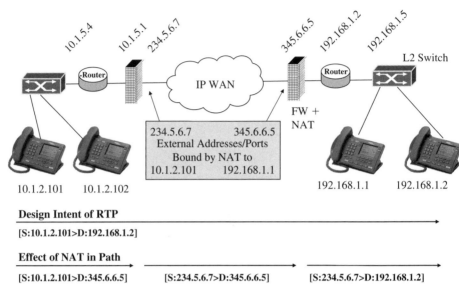

FIGURE 8.4 Voice packet addressing through NAT at both ends.

Call control servers assign UDP ports for RTP media packets of new call connections on a random basis, from the range of port numbers above both the "well-known" and "assigned" ranges. A firewall normally closes all those ports, blocking packets addressed to them. A smart firewall or application layer gateway (ALG) can monitor the fixed UDP or TCP ports for signaling, read the traffic to learn when UDP ports are assigned to new calls and when calls terminate, to open "pinholes" for UDP ports only as needed. Most firewalls are not that smart. Again, NAT complicates the procedure enough to break the basic protocols.

A further complication arises from the original expectation that the RTCP (control) session for a call will be assigned the next higher port number (odd), above the RTP port (even) for the call itself. NAT devices don't necessarily maintain this relationship, and seldom can reserve the next higher port when opening a port for a voice call. For this reason the IETF developed RFC 5761 to standardize a way for RTP and RTCP to use the same UDP port, simplifying the NAT problem by needing only one port opened. The former requirement that RTP occupy an even port number and RTCP occupy the next higher port no longer applies.

A way around the problem of NAT and firewalls is to put a combination proxy/registrar outside of them. When a UA registers from the inside of the firewall, it will open a port for that connection since it was initiated from inside, and most NATs normally behaves this way. By keeping the connection open, these proxies can use it to forward messages to the UA, for example, an INV from a caller on the Internet. Recommended practice is to maintain an encrypted TCP connection from the firewall to the outside SIP proxy, with TLS, and use it for all such traffic. Rather than have each UA register with a server outside the firewall, the registrar can be inside and open a shared TLS connection to the outside server. See Section 8.2 on SIP Trunking for more information.

IPv6 was supposed to eliminate the need for NAT: there would be plenty of addresses for everything. Security issues, however, favor NAT as a way to hide from outsiders the topology of a private network. Today the "best practice" is to assign private or unroutable IP addresses to VoIP devices so that they can't connect directly to (or be attacked from) the Internet—making NAT necessary indefinitely.

8.3.1 Failures of SIP, SDP (Signaling)

NAT installations with firewalls commonly require that all connections originate "inside" the NAT function (inside the firewall which supports NAT). However, a signaling message for an incoming call, by definition, wants to originate a session from "outside." There are ways to allow incoming calls, for example, by having the internal call processing server set up a persistent connection, through the firewall and the NAT function, to an Address Of Record proxy server outside. You can also place a call proxy server on a public address in a DMZ.

A public proxy outside the firewall still needs the UA client or its proxy to set up a connection so incoming requests can pass the firewall. The client does this using a REGISTER request to a proxy address. The UA may learn the proxy's address from DHCP, for example. A device "locked" to a carrier could have the address of the carrier's host server imaged into the UA's firmware.

For signaling, the transport is TCP. The client keeps the connection alive, and preserves the address-port bindings in the NAT box, by sending keep-alive messages every minute.

UA clients can set up multiple connections, or "flows," to a single or multiple external proxies. If a connection is lost, the client can open it again with a re-INV message using the same "reg-id" parameter in the Contact: header.

A SIP proxy server inside the firewall can register all the IP phones and terminals on the LAN, establish a shared connection to an external proxy (or two, for redundancy), and thereby reduce the complexity of end point devices. That shared connection between internal and external servers will be easy to keep alive, perhaps from normal traffic with little need for keep-alive messages (except over night). The path is open to allow all SIP packets to pass the firewall.

8.3.2 Failures of RTP (Media)

Media transfer of voice and video over IP was designed to take place directly between the end points over RTP/UDP/IP (Figure 8.4). That is, the destination

IP address inserted into a voice-carrying IP packet by the speaker's IP phone should be the IP address of the listener's IP phone. The UDP port number is assigned by the call processing server during call setup.

In general, a calling device learns the address of the called device from a call processing server. However, call servers inside the firewall must rely on the local host addresses (which may be private 10.x.x.x format) of the end points they serve. Private addresses are never valid outside their own domain. Where NAT applies, even a routable inside address is not usable by an external peer. Further, where private addresses are assigned to IP phones, there may be conflicts or overlaps of address ranges assigned at both sites, a perfectly legitimate practice.

Rather than the destination phone's host IP address, the originator of an RTP connection needs to know the public IP address on the NAT device that corresponds to the (private) IP address of the called IP phone. This is also known as the "reflexive address," "reflexive server address," or something like "reflexive UDP admission." The receiver's site NAT then maps the public WAN IP address of the called IP phone to its local or private IP. Both ends of a call may also need to map the UDP port numbers through port address translation (PAT, making it NAPT) as well as NAT.

8.3.3 Solutions

The solution to the problem requires a method (protocol) to inform an IP phone of its own public socket addresses (IP + UDP port):

- Assigned to itself by its own NAT.
- Which the far-end NAT will translate into a private address on the LAN of the targeted IP phone.

Vendors offer a variety of solutions based on services and proxy servers that may be based within the enterprise or operated by a carrier. These servers may be applied in combination, depending on the network topology.

The goal is to enable the VoIP end points to reach the signaling server, then to communicate with each other. To do that, the signaling server must also be able to send messages to the end points. There's an RFC for that: 5626. The scenarios resemble these:

- UACs inside a firewall/NAT use a home proxy on the outside. Each UA end point maintains a registration connection to a UA in its proxy. The two UAs exchange messages (via one or more proxies) on the persistent "flow" that each maintains through the FW. Calls from the Internet reach through the FW/NAT to the UAS for an end user on the same flow. See the description of STUN.
- UAs behind the FW/NAT bind to a public IP address on a server that relays messages through the FW on connections that originate from the UA. This is TURN.

• Interactive Connectivity Establishment (ICE) is a procedure that combines STUN and TURN. It has wider applicability than VoIP or SIP, and usually works, even when both of the other two methods fail to allow connections.

8.3.4   STUN: Session Traversal Utilities for NAT

STUN formerly stood for Simple Traversal of UDP through NATs, but it was expanded to include TCP and other protocols in a new RFC (5389). The original version (RFC 3489) didn't work well enough in practice and suffered from security vulnerabilities. The new name reflects its broader applicability to protocols other than UDP, in particular, TURN and ICE (see below). There's also the fact that the original STUN by itself didn't work well in all real-world situations.

One STUN component resides on a server that has an IP address in the public Internet, typically, or in a network segment where all interested parties can reach it. The other component is a client that queries the server from behind a NAT firewall to learn what public socket (IP + port) the NAT device assigned to the client's private address.

IP addresses of STUN servers may be configured into network elements such as the hosts file on a PC. Proper DNS record entries, under the domain of the enterprise or the service provider, can also let clients discover the address for STUN service.

In operation, a STUN server receives a query from a STUN-enabled end point located on the inside of a NAT process. The message format is shown in Figure 8.5.

The only method defined originally is "binding," which relates to the mapping of the private socket to the public socket—they are said to make up a binding. Attributes have codes assigned by IANA and listed in a STUN registry.

If STUN operates on a TCP connection identified from a DNS query, using its port number of 3478 or 5349 as assigned by IANA, there is no need for additional framing or protocol encapsulations. Other use cases require additional headers, for example, when STUN shares a port address with other protocols or the TCP connection is encrypted with TLS (also called SSL). Three features of STUN help identify its packets:

• The first two bits of the STUN header are always 00.
• The length field (in octets 3 and 4) represents a multiple of 4, so it always ends in binary 00.
• The magic cookie content is fixed for STUN packets.

STUN attributes are the payload of the protocol. They convey the address information and other factors to authenticate the message or explain use cases.

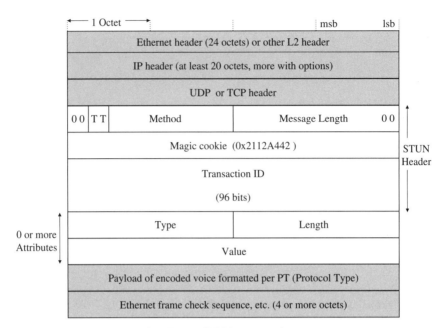

FIGURE 8.5 STUN message format.

- **Mapped-address:** the public socket of the device making a query. This response could be either IPv4 or IPv6 regardless of the private address version.
- **XOR-mapped-address:** the public socket obfuscated by the XOR process, for privacy. Used if the magic cookie is present.
- **Username:** the name and password of the requesting device.
- **Message-integrity:** the SHA hash of the STUN message from the header to the attribute before this one. Placed last except for the Fingerprint. Use of this attribute reduces the threat of denial of service attacks via STUN.
- **Fingerprint:** a CRC-32 check sequence form the header to just before this attribute field.
- **Error-code:** appears in responses that report errors. Contains the numeric code (300−699) and a UTF-8 text phrase describing the reason.
- **Realm:** globally unique identifier in human-readable form, typically a domain and TLD. Presence of the Realm attribute indicates a long-term credential is in use.
- **Nonce:** number used ONCE, a generated number to identify a transaction.
- **Unknown-attributes:** fields of 2 octets each (from the IANA registry) report the identity of attributes in the request that the server doesn't understand.

- **Software:** the UTF-8 coded text describing the client or server vendor and version.
- **Alternate-server:** redirection to another site if the addressed STUN server can't handle the query.

When the query packet from inside the firewall arrives at the STUN server, its header contains as the source address the translated or public IP address and port that was inserted by the NAT (Figure 8.6). This socket replaces the host IP and port for the originating end point. This socket cannot be used by the called phone because an inside address is valid only behind the firewall. The STUN server returns this socket (IP + port), the reflexive address, in the payload of the response. From that the asking end point learns its public socket.

When negotiating a connection with a device outside the firewall, the end point can now give its public address, to the called party, not its actual host address on the private network. The called end point responds to the call by sending packets to its public address of the calling end point.

As described, classic STUN has a problem with generic application level gateways (ALGs) built into some firewalls and NAT boxes. The ALG looks for IP addresses deep in each packet. If it finds any and there is a NAT binding for it, the ALG rewrites the IP address before the packet goes further. At one time Cisco routers defaulted with "SIP Fixup" on. Names for this function at other vendors include SIP Inspection and SIP Transformation.

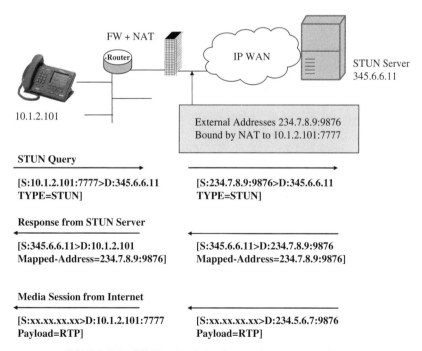

FIGURE 8.6 STUN network topology and message exchange.

The STUN update in RFC 5389 encodes IP addresses, to hide them from generic ALGs. You may find that these functions in NAT that interfere with STUN can be turned off in most devices.

STUN is not effective with symmetrical NAT, or in cases where both ends are behind different NAT boxes. As a result STUN sees use as part of a process rather than as a stand-alone feature. How to use STUN is described in a usage document, for example, the descriptions of TURN and ICE, which follow.

Note that the STUN server provides only an address discovery service—it is not in the path of either signaling or media packets. A call control processor or SIP proxy is necessary for devices to register their locations and find each other. An attacker can interfere with STUN by changing a return message in transit, which causes it to be dropped and prevents the transaction from completing.

A distributed DoS attack is possible by spoofing the source address in a query, inserting the IP + port of the targeted victim. Every query generates a message to the target. Filtering input message addresses can protect against this form of attack.

8.3.5 TURN: Traversal Using Relays around NAT

To work with symmetrical NAT, or to handle NAT at both ends of a connection, a TURN server located outside the corporate firewall (or at a carrier) relays both signaling and media packets. A TURN proxy accepts packets from one device and forwards them to the other. In the process, TURN changes packet addresses, recalculates CRCs, and may reset fields such as time to live (TTL).

Designed to work within ICE (see Section 8.3.6), TURN also can operate alone. However, clients need a mechanism outside of STUN and TURN to learn each other's public sockets on the TURN server. The RFC suggests email, but there are automated methods, possibly DHCP. How the clients find the TURN servers also is part of ICE (see below).

To set up a connection, both end devices or clients must request an address binding from the same server or servers that can locate each other. The client sends an ALLOCATE message to the server (Figure 8.7). The TURN server allocates a unique public socket on itself and responds to the client with that information in a AllocationConfirm message.

The full title of the TURN RFC states it is relay extensions to STUN. The response from TURN contains information similar to that provided by STUN, so a network doesn't need both. The difference is that STUN servers return a socket on the client's NAT; TURN identifies a public socket on itself.

This description of TURN assumes, as in Section 8.3.4 on STUN, that the messages used by STUN are also used by TURN, and in the same formats with the exception of ChannelData. So that discussion will not be repeated in detail.

Key extensions in TURN are additional client message formats. In descriptions the "client" originates a connection to a "peer." Each client or peer must ask the TURN server to allocate a public socket for its use.

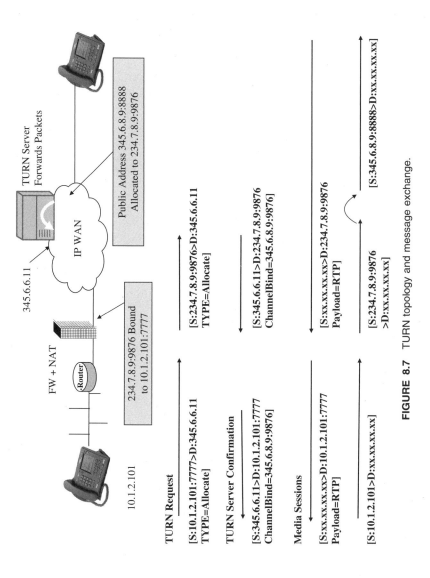

FIGURE 8.7 TURN topology and message exchange.

- **Allocate Request:** client asks for a public socket on the TURN server.
- **Allocate Refresh:** Client keeps the allocation alive by extending its expiration time.
- **CreatePermission:** client gives TURN server permission to forward data to and from another client's public socket.
- **ChannelBind:** allocates or refreshes a channel number and binds it to a connection, a peer's public socket; also creates a permission for the connection.
- **DataIndication:** packet from TURN server to client.
- **SendIndication:** original data transmission from client to TURN server.
- **ChannelData:** send message with a 4-octet header (a 16-bit channel number and a length field) replacing the 36 octets in a STUN-formated header.

A client may format outbound data as a SendIndication or ChannelData. The server will always use ChannelData if a channel is bound to the connection. This reduces the bandwidth requirement.

One IP phone sends packets to a second IP phone by addressing the public socket address of the second phone, which is on the TURN server. The server translates addresses and forwards packets to the NAT function at the second site. The second phone uses the same process, addressing the first phone's public socket on the TURN server. Each phone sees the server's address in source fields of headers in received packets.

8.3.6 ICE: Interactive Connectivity Establishment

The IETF document *Interactive Connectivity Establishment (ICE): A Protocol for Network Address Translator (NAT) Traversal for Offer/Answer Protocols* (RFC 5245) applies to many protocols of this type. ICE allows two UAs to learn public addresses or sockets where they can exchange media packets directly between end points by the shortest or lowest cost path. Both end devices must support ICE for the procedure to succeed.

SIP signaling paths between UAs and their authoritative proxies outside the firewall are set up via ICE servers. ICE uses the Session Traversal Utilities for NAT (STUN) Protocol and its extension, Traversal Using Relay NAT (TURN) to allow the UAs to exchange SIP messages (see Section 5.2.1 on SIP Architecture).The end points inform each other where they can send media packets by including addresses in the Session Description Protocol (SDP) blocks in the SIP packets.

The idea of ICE is for each UA to give the other a series of IP + port (socket) addresses (via SDP), with a priority for each (called the preference, as q= some value between 0 and 1). Each UA then tries to connect to each of the addresses, in order of preference, until one works or the list is exhausted. When each end finds an address that works, they then have peer-to-peer connectivity for media sessions.

The ICE process starts when each UA registers its AOR via SIP with a proxy server at a public IP address. The UAs may be behind NATs, which means that they could have used STUN and TURN to register, from which each UA would know several possible addresses for itself. Each address represents a potential path for direct communications.

Addresses would be a unique "transport address" or socket, an IP address and a UDP port, which may be different for RTP and RTCP (but can also be the same). The recipient of signaling packets on ports open for both RTP and RTCP can distinguish between the two protocols by the protocol type.

CANDIDATES are drawn from addresses such as:

- A directly attached network interface ("Host" address; Ethernet, Wi-Fi, VPN, etc.).
- A translated address on the public side of a NAT ("server reflexive" address).
- The address allocated from a TURN server ("relayed address").

Each UA offers a ranked list of "candidate" addresses (the CHECK LIST) in an SDP block of a SIP Offer message. To test for connectivity to these addresses, both ends then send packets directly to each other's sockets (IP + port) designated for media packets. These packets are not sent to a STUN server, and are sent to a TURN server only if more direct paths are not possible. Test packets originating at each end open ports in the local firewall, pinholes that any NAT will find acceptable for returning messages from that candidate address.

When the reverse check gets through, the path is opened end to end in both directions without communicating anything special to the NAT devices and without having a SIP-aware firewall. ICE works regardless of the type of firewall, which can map and filter with or without a dependency on the external address or port.

Keeping the ports open is an additional consideration. The process depends on the NAT pinhole remaining open while both UAs send the test messages. The STUN keep-alive is used. The typical default inactivity time out for open ports, 30 seconds, should be enough. The timer value can be increased to something like 3 minutes so the frequency of the keep-alive messages can be reduced to save bandwidth and processing cycles.

8.4 SESSION BORDER CONTROLLER

The point where two different IP networks (under separate administrative control) exchange voice traffic might not be unique. Tier 1 Internet service providers may exchange traffic at multiple geographically-separated sites to improve availability, minimize path lengths and latency, or balance loads. However, if a carrier wants to manage and track voice calls for the purpose of billing or to more easily identify packets that should be prioritized, then those connections are routed to designated interchange points specifically designed for voice.

The session border controller (SBC) is purpose-built hardware and software to deal with the processing-intensive tasks of handling voice traffic. Its functions include:

- Emulate and replace a firewall:
    - Application aware SIP firewall, open/close UDP ports based on SIP signaling messages.
    - Translate network IP addresses (NAT) and ports (PAT); together, Network Address and Port Translation (NAPT).
- Transcode voice packet payloads to/from PCM, ADPCM, G.729, and so forth.
- Transcode video stream formats.
- Translate signaling formats between SIP and, for example, H.323, SIGTRAN, ISDN, channel-associated signaling (CAS), or SS7.
- Control the admission of new calls to prevent denial of service attacks.
- Bridge UDP connections to TCP (Microsoft uses TCP for voice packets).
- Recognize and properly handle various media types: voice, fax, video, data.
- Mark media packets for proper QoS handling in the network.
- Generate billing records.
- Protect privacy via encryption of voice streams and signaling messages. For example, encrypt RTP from the LAN to sRTP on the WAN side.
- Register and authenticate users.
- Focus troubleshooting and network analysis.
- Create duplicate packets for a wire tap or call recording.
- Monitor for problems and issue alerts regarding jitter, dropped packets, and latency.
- Bridge RTP traffic among VPNs; the SBC joins each VPN and proxies all end points in all VPNs.

Two SBC markets have developed: enterprise and carrier. Functions are similar for both, but speed and capacity must be higher for the carrier environment. SBCs come in sizes intended for a small to medium business (SMB) to beyond 50,000 concurrent sessions for carriers. Currently the capacity ranges up to 10 Gbit/s. SBCs are behind SIP trunking services.

The functions of an SBC require them to be in-line of both signaling and media streams, hence the high-speed interfaces.

8.4.1 Enterprise SBC

A common architecture for the SBC is back-to-back User Agents (B2BUA) (Figure 8.8). A Registrar service in the same device simplifies network design and can off-load record keeping to a backend RADIUS database (DB) server on the LAN or draw on an existing database such as one kept by Human Resources.

SBCs are mostly considered for the position between IP networks, but some offer integral MGWs. That function logically is separate and is covered elsewhere in this book.

The enterprise SBC has separate interfaces to the public and private networks. All voice and video traffic between the two networks pass through the SBC:

- One interface handles all calls between the user-side IP phones, gateways, cameras, and other end points to the SIP trunks. These devices may sit behind other NATs and firewalls inside the LAN, and often have private IP addresses.
- The other UA has a public IP on the Internet, the registered and published location for all the phones and terminals on the private side.

Because it sees and handles all the packets, an SBC can exercise extensive control over traffic. Under some circumstances the SBC may step out of the path for media while retaining the signaling path.

When a call request arrives from one side, the SBC terminates the call on a proxy UA. Before it generates a new INVITE message from the other side the SBC may demand authentication credentials or confirm the validity of the called address in a database (DB).

The transcoder/converter function between the UAs adjusts the headers of packets as they pass through, including NAT. This behavior hides the topology of the enterprise network from the Internet.

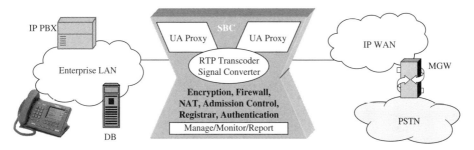

FIGURE 8.8 Enterprise session border controller, which sits between a private LAN and SIP trunks to an IP carrier and public networks.

To add value, an SBC may apply Intrusion Detection/Prevention, threat analysis, and other security measures to the traffic between the two UAs.

8.4.2 Carrier SBC

Carriers like the proxy functions of SBCs because they allow all voice traffic to funnel through there, the RTP voice packets (the bearer channels) as well as signaling (SIP or H.323) (Figure 8.9). Confining voice to a physical path simplifies tracking usage for billing records and troubleshooting audio problems. Path control also allows network engineering to provide high priority for voice packets and creates a point where Quality of Service (QoS) can be observed. With a focus point for VoIP, the carrier can tap a call in fulfillment of its obligation under CALEA.

By observing the SIP message headers, an SBC can determine what kind of NAT firewall, if any, guards each end point. If the end points on a call have public addresses (sockets) on the Internet, or are mapped to one by a not-too-strict NAT firewall, the SBC can issue re-INV or REFER messages to both ends after call setup, telling them to send media packets directly to each other.

If either end point is behind a port-mapping symmetric NAT, the SBC will have to stay on the media path. Those NAT's are too restrictive to pass packets with addresses (sockets) that result from a call transfer or referral command to start direct packet exchange. The source addresses in the arriving packets are not the same as the destination addresses of the client-originated packets that opened the pinholes in the firewall. Continuing to proxy the media streams is not optimum. The SBC in general may be on the shortest allowed path (when voice is focused there), but there may be paths with fewer hops. An SBC needs time to process a packet through two UAs, so it will add some latency.

Software vendors offer application to monitor revenue and even margins, based on billing rates. Real-time control allows carriers to route voice calls to maximize profit. Enterprises can use the same features to route calls over the

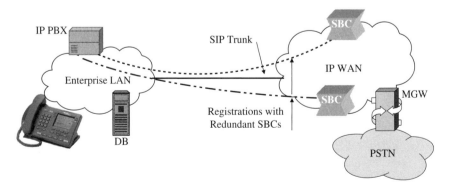

FIGURE 8.9 Carrier's session border controller (SBC) serving enterprise phones.

least expensive path or to ensure that users hear the highest available voice quality.

The SBC hardware platform may incorporate other security measures such as access management that requires authentication of callers or management of link bandwidth. SIP trunking services may use an SBC to authenticate customer IP PBX's and ensure they register the addresses they proxy.

For the enterprise, the carrier's SBC is beyond the point where the enterprise edge router exchanges all VoIP traffic. The carrier's SBC IP address typically is statically configured into the call control server, firewall, and other enterprise network equipment. An enterprise might choose to place its own SBC here (Figure 8.10) for all of the reasons bulleted above. In addition a separate path for voice and UC sessions relieves the data firewall of considerable processing while adding voice-aware functions that might be missing from a traditional firewall.

That path to the carrier can be a dedicated T-1/E-1/PRI, a TDM channel in a larger pipe (e.g., SONET), or a logical allocation of capacity such as a VPN on an OC-3 or similar IP connection. On the "inside" of an enterprise SBC, look for a gigabit Ethernet port, preferably at least two.

Whether there is one SBC or a pair (carrier and enterprise) makes no difference to users. The SBCs forward traffic to each other transparently to callers.

Two SBCs (enterprise and carrier) can simplify operations. All the VoIP devices within the enterprise operate basic software and standard protocols with no extensions required. All transmission can rely on private IP addresses without encountering any difficulties from NAT or needing STUN, TURN, or ICE servers. The link between the two SBCs can be configured statically just once for all VoIP traffic, reducing the need for keep-alive messages. One path lends itself to economical encryption, simplified provisioning of good QoS and traffic engineering, and easily designed redundancy.

The carrier's SBC registers the phones and publishes (in the DNS) its own Internet address as the location to contact the enterprise domains for VoIP

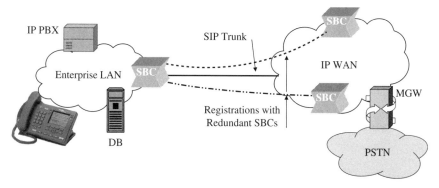

FIGURE 8.10 SBCs in both carrier and enterprise networks. The enterprise SBC may be doubled for redundancy too.

service. There can be a drawback with a single external proxy server when the two parties to a call are behind the same NAT firewall. In the simplest configuration, where both phones register with the external proxy, the media packets will be routed out and back through the firewall. Another advantage of an enterprise SBC is that such calls are kept on the LAN. The CPE SBC should perform "hairpinning" to route packets out the same physical Ethernet connection on which they arrived, preventing the double passage through the firewall. Routing policy defines this behavior when both phones have addresses on the LAN. Draft IETF documents show interest in finding a way to avoid hairpinning.

8.5 SUPPORTING MULTIPLE-CARRIER CONNECTIONS

A goal of achieving the highest possible availability can justify redundant local loops. However, dual access won't gain much if they both access your premises at the same point and on the same physical cable. It can happen easily if two local loops are leased from the same LEC. Diversity must be complete to reap the most benefit. This means separate entry points, preferably at opposite ends of a building. Each access line should terminate in a different CO.

Sometimes its necessary to trace the physical route of those cables, to ensure that they don't intersect. One report describes how two access lines followed separate aerial routes, except where they crossed—both cables were mounted on the same pole. A truck took down that pole.

If your PSTN access is from a SIP system via a media gateway, the installation is relatively simple. On the LAN side, all gateways look the same in providing access to global telephone numbers. On the PSTN side, the physical ports are split between access links.

For greater assurance of availability, use access circuits from two carriers. Two SIP trunking providers should assign DID numbers from different ranges. They will need an interchange agreement so if your access link from one SIP provider fails you can receive those calls on the other access.

Deploying Internet access to more than one ISP (Figure 8.10) gets more complicated. The enterprise SBC will register with each ISP. Both ISPs will advertise their IP addresses as the location of a SIP proxy. Incoming calls will find one or the other ISP, depending on how the caller's DNS query resolves.

The enterprise's edge router will need to participate in the Border Gateway Protocol (BGP), the Internet's way of finding autonomous systems (ASs) that represent a management domain such as a carrier or enterprise. If you connect to only one ISP, it is the destination AS from the Internet; the customer doesn't need an AS number. When the enterprise is a customer of multiple ISP's, the enterprise network needs an AS number assignment (from IANA) so it can be found via either ISP.

Under normal conditions, an enterprise typically balances the load between carriers by reporting a different subset of its IP addresses to each carrier's edge router. If one Internet path fails, BGP automatically moves all traffic to the other.

For optimum voice quality, IP management appliances can monitor the packet loss and latency of IP carriers. This QoS information is fed back into the enterprise router to adjust the routing and forwarding tables so that outbound packets take the network that is performing better at that time.

High-availability arrangements will evolve. Check for new capabilities before deciding.

8.6 MOBILITY AND WIRELESS ACCESS

8.6.1 VoIP on Wireless LANs/Wi-Fi

Some mobile handsets offer the user multiple options to place a call on:

- The cellular service.
- A VoIP connection on a local data network, typically an 802.11 wireless LAN (Wi-Fi).
- A VoIP connection on a mobile data network such as EV-DO, LTE, and WiMAX.

In early implementations, the two options were not coordinated. That is:

- Calls on the cellular network counted against a cellular service plan's minutes, was routed by the carrier, and had access to related services such as call recording or voicemail.
- A Wi-Fi call came from a VoIP client application running on the phone and relied, typically, on a SIP server provided by the enterprise or a third party. It was not related to the cellular service.

A more unified approach has developed that integrates the access methods under the cellular carrier's call control. In this case the user doesn't have to choose the access path, and might not be aware of the selection. All service features are available at all times. The most sophisticated services move from one connection to the other, depending on the presence of a Wi-Fi access point or cell tower, relative signal strengths, or an evaluation of traffic loads.

With the 4G cellular network (LTE and WiMAX), the phone function will use a packet data service for all calls. Every connection on the radio interface is packetized.

Today voice can share the radio link when the data rate is high enough. The carrier may choose to host the VoIP/SIP call control server or partner with another provider. Business models will evolve for some years as new carriers attempt to attract customers with novel pricing plans.

In the dynamic arena of cellular, rules and plans change constantly. Some carriers won't allow VoIP clients (SIP UAs) to place calls on their mobile data services. Or they charge extra per month to use an LTE smartphone as a Wi-Fi access point for a LAN device (which might be a Wi-Fi phone). There are charges that such restrictions violate the terms of the carrier's FCC license for the radio spectrum.

8.6.2 Integration of Wi-Fi and Cellular Services

A smart phone keeps track of what connectivity options it has at all times: what cell tower sites are around, how far away they are, how much transmit power is needed to reach them, what interference exits, and so on. A growing number of phone handsets also carry a Wi-Fi transceiver to make the same survey of nearby hot spots.

With that knowledge, the mobile unit can decide which form of RF signal to use: cellular, mobile data, or wireless LAN. The feature appeared first from PBX manufacturers whose mobile handsets could hand off a conversation in progress from one form of connectivity to the other, either way. For example, a person might receive a call on the mobile while at his desk, then leave the building while continuing to talk. The PBX would hand off the call to the cellular network to pick up when out of range of the Wi-Fi access points in the building.

Large carriers with both types of network have offered mobile handsets that work in similar fashion, but may be limited to Wi-Fi hot spots operated by that carrier. The benefit to the cellular company is reduced load on the cell tower as traffic is moved to a Wi-Fi hotspot that possibly is on another company's network. Multiple radio interfaces and the ability to hand off calls among them are important consideration as smart phones deliver more video and web surfing that drive up demand for air-side capacity.

Upstream from the cell tower is another congestion point, the backhaul from there to the switching center. Originally the need was for voice channels, satisfied by one or more T-1 lines from the local exchange carrier. But as demand increased, with data exceeding voice volume, more capacity came from direct fiber links where possible, microwave links, and carrier Ethernet on faster and faster infrastructure.

8.6.3 Packet Voice on Mobile Broadband: WiMAX, LTE

Applications such as SIP user agents have started to appear for mobile phones. The software allows these phones to use the data service for VoIP rather than the switched cellular voice service. Some cellular carriers resisted, and not all have yet accepted this mode of voice. It seems inevitable, with the barrier is breeched, that VoIP on mobile BB data connections will become a widely used service.

Into 2011 the cellular carriers had not settled on a strategy for handling voice over broadband wireless networks. Reluctantly, they allowed a Skype application on smart handsets so calls could be placed using the "data" capacity of the wireless link to the base station. These calls count against the data cap, if there is one, but don't use "cellular minutes."

Going forward, a new standard is expected for the all-IP fourth generation (4G) broadband cellular services. In particular, the Voice over LTE (VoLTE) approach is seeing progress in setting standards and talking about adoption by major carriers.

8.6.4 Radio over VoIP

Handheld radios provide push-to-talk voice connections for police, firemen, and managers of the public forests among other government employees. Construction workers, plant managers, and others in the private sector depend on the radios as well. Transceivers and repeaters extend the range available directly between the handhelds. Like cell sites, radio repeaters work best when their locations have some height above the average terrain. Often they are in remote areas, or temporary, but still need back-haul connections to head-quarters, typically radio links and phone lines.

Organizations with IP networks may have connectivity for part or all of the way. Hence the practice of "radio over IP." They extend the range by linking repeater transceivers across an IP network.

In one example, an IP network between two media gateways connects to handhelds. They connect to the MGWs using their 4-wire E&M interfaces (Figure 8.11). The E&M is the least common denominator interface of the telephone industry, supported by voice switches, PBXs, and transmission equipment for almost 100 years. Handheld radios use the separate pairs for earphone and microphone (the 4-wire part of the name) on the voice path.

SIP call control allows one MGW to locate the other and set up the con-nection. A transmission from a radio within range of either radio on the IP network will be picked up and relayed to the far end of the IP connection. That

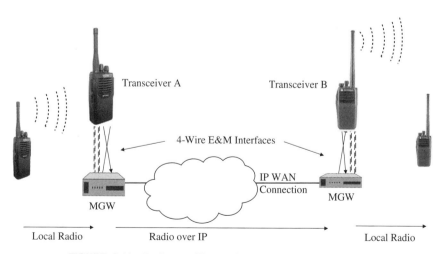

FIGURE 8.11 Radio over IP encodes analog interfaces as VoIP.

radio will rebroadcast the speaker's voice to radios in its vicinity. The E&M signaling state tells the MGW when to apply silence suppression to prevent sending packets if no one is speaking on a radio.

The assertion of the M lead by the radio affects the M lead on the far-end MGW, which "pushes" the E lead button to turn on the transmitter.

Any form of IP connectivity works. Routing the IP traffic over a satellite extends mobility to most of the Earth's surface. A small ground station provides access for the remote radio. The extra latency of the satellite circuit hardly matters because the half-duplex conversation is controlled by push-to-talk.

IN SHORT: E&M Voice Signaling

E&M is one of the original methods, from a time when the only signaling other than the POTS line was between switches. E&M illustrates signaling at its most fundamental, and perhaps this is why it remains in use so long after it appeared on the first automated switches.

In modern terms, E&M separates the control plane (E and M signaling leads) from the message plane (copper analog voice paths). The interface lies between two peer devices that want to set up and end connections on the voice path (Figure 8.12).

Inside the PSTN, most voice paths are 4-wire: a separate UTP for each direction. E&M also operates with a 2-wire voice path, something like a loop-start trunk.

The calling side of the interface signals a trunk seizure by asserting it M lead, which does something to the other side's E terminal. The mnemonic is that the Mouth speaks to the Ear. In England they think of Earth (a current sink) receiving a stimulus from Magneto (an electrical generator).

Exactly what "assert" means defines the five types of E&M signaling. The power source may be on either side. Assertion includes applying power or grounding a lead. Most equipment supports more than one type (Cisco doesn't support type IV, not much used any more). The details are easily found and will not be repeated here.

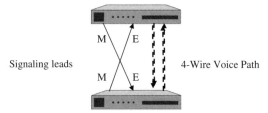

FIGURE 8.12 Each side signaling the other across an E&M interface by "asserting" its own M lead, which affects the E lead on the peer device.

9

NETWORK
MANAGEMENT
FOR VoIP AND UC

Organizational convergence follows the unification of voice and data. When all services share one infrastructure, they almost necessarily share management and troubleshooting as well as transport. There may be exceptions in larger organizations where it is possible to dedicate specialists to an area such as troubleshooting voice quality issues. More on this later.

The choice of a network management system (NMS) can have a surprising impact on a business. In early 2011 Nemertes Research studied the costs for VoIP management. They found that a specialized management tool for VoIP, one that measured network performance attributes that correlate with voice quality, reduced the cost per seat by more than half compared to using no NMS. The cost when using the vendor's product-specific workstation fell in between.

The cost differences derive from the time needed to solve problems. A specialized tool that addresses all devices and all protocol levels finds faults faster. Less labor equals less cost equals reduced staffing.

Unexpectedly, having more than one specialized manager of managers (MoMs) was worse than having none. The analyst's explanation claims that a "single pane of glass" for management is the key to cost reductions. Multiple panes incur multiplies costs.

One pane to watch implies that one person will do the watching. By implication, organizations that formerly had separate telephone and data staff must

VoIP and Unified Communications: Internet Telephony and the Future Voice Network, First Edition. William A. Flanagan
© 2012 John Wiley & Sons, Inc. Published 2012 by John Wiley & Sons, Inc.

cross-train everyone: that watcher of the pane must know all aspects of the network:

- Voice switching and signaling, QoS analysis.
- Layer 2 data switches and L3 routing.
- Transmission technologies in the network.
- Carrier access and services: SIP trunks, SONET, Ethernet.
- Messaging services and servers.
- Call center operations.
- Call control protocols (signaling).

No more silos. Management must take an end-to-end (unified) view. It is necessary to have visibility over all vendors for faults, performance metrics, and configuration management if the system is to correlate multiple responses to find the root cause. To be most effective, automation aids and tools need to cover all brands of equipment, software versions, services, and applications. Finding a "where" for a voice problem is tough and requires visibility to key metrics of each platform or service.

The use of such a tool, particularly for configuration management, contributes to good governance at all stages from planning through deployment to operation. A budget for an NMS should anticipate growth in traffic which will require more:

- Cross training.
- Backbone hardware.
- License fees.

However, reduced staffing could lower costs. Based on the savings from a multifunction NMS, it deserves a separate proposal.

Trend reports help isolate faults, predict problems. Dashboard displays in the NOC can show clearly key metrics or calculated (compound) metrics. A complete NMS will capture and report on a poor-quality call while it is still up, perhaps catching transient errors that are not visible after the call completes.

9.1 STARTING RIGHT

Wherever you are in deployment, a best practice is to baseline the network. Know what you have before you go further. It's amazing how much a network changes, and how much can be added, without being precisely known.

Collect baseline data before issuing the RFP for the project. Look at two years if the data are at hand. If you are not preparing the baseline with current in-house staff, the contractor who does the initial assessment (and perhaps works on the RFP) should not be a vendor who will bid on the installation or maintenance tasks.

If voice is not yet migrated to IP, or has not progressed much, a pre-assessment is most important for success with VoIP/UC. Major vendors insist on evaluating an existing IP network before adding voice to it—they don't want a failure they can avoid. You don't either.

The automated discovery performed by various commercial software products minimizes the labor involved in creating the "as-is" map of the network. There are many commercial and open source applications to discover a network and map it. With some planning, the collected details of the topology and installed equipment will transfer to other software tools that monitor and manage the infrastructure after it goes into production.

Planning for deployment must consider the need to configure customer equipment to be compatible with network services. Who will perform this work isn't obvious at all. A SIP trunk provider of proxy, STUN, TURN, and ICE services probably won't set up customer equipment connected to another carrier's line. Will the LEC? Will the alternative access provider who brings in a fiber? Will the long-distance carrier? Sounds like you might need an RFI before the RFP.

After the installers leave, how will you maintain the equipment and services? Best to get all the manuals and detailed specifications very early in the process. In particular, get the configuration guides that point out the known problems and workarounds for the software releases and hardware version you buy. Results from interoperability tests are important too.

9.1.1 Acceptance Testing

The RFP and proposal remain important until the VoIP/UC deployment is accepted. The RFP or Statement of Work must include an acceptance test plan. It must be detailed and inclusive enough to confirm all of the requirements in the RFP that the vendor accepted.

Some Service Level Agreements (SLAs) such as percent uptime are hard or impossible to test at the start; they require measurements over an extended time. Other SLAs such as maximum latency, minimum throughput, and call setup time can be tested at any point.

It is also important to confirm the operation of all features while executing the test plan. If end users expect to be able to set up ad hoc conferences for 12 callers at any time, this must be confirmed, as must the number of separate conference calls. Similarly for the number of simultaneous calls on a trunk, after-dial delay, and information displayed on IP phone screens. Don't neglect a fax transmission.

Document everything. Get it in writing (or on a memory module). Yeah, websites are forever, but you might be thankful for a paper copy of a software manual when rebuilding a downed server after the tornado/flood and before Internet access is back.

9.1.2 Configuration Management and Governance

Configuration management (CM) is an ongoing effort, starting with the baseline configuration recorded before starting a new project. Large organizations see CM as part of Enterprise Architecture (EA), the task that inventories and models not only hardware, software, software versions (which really matter), and networks transmission links but also the business processes they support. Continuous monitoring updates the data base or model to reflect the "as-installed" state of infrastructure.

CM is more than backup files of device configurations, or centralizing the inventory of hardware and communications lines. CM controls and manages change through defined business procedures that require management approval that's separate from a requester before anybody makes a change.

COTS software exists for CM and EA. Certain high-end management systems include enough CM capability for many situations. Top managers and auditors will want to track changes in detail: when, on what, by whom, why, where, and so forth. The best CM software doesn't allow edits of the files, only additions to correct previous errors, to protect the audit trail.

Strict organizations, including federal financial and military agencies, don't allow a new device or application on their servers or networks unless reviewed, vetted, and approved in advance. Plugging in a unknown device in these offices brings a swift response.

A governance body backed by a strong policy on CM will manage the constant monitoring of the network. Awareness of unauthorized devices and changes offers protection against the introduction of rogue wireless access points, personal laptops, or other vulnerabilities.

9.1.3 Privilege Setting

An important tool to secure a network or VoIP system is the range of permissions the root administrator can grant to other admins and to end users. Each phone, and each line appearance on a multiline phone, can have different privileges regarding hours of operation, access to long-distance service, international calling, encryption, priority of access to resources such as conferencing, use of video and fax, and many others.

For small networks the admin or a contractor may configure phones manually. This procedure becomes impossible for even moderate size installations. Automation aids are necessary for most enterprises:

- Firmware updates by downloading from an on-site server. Usually done automatically when an IP phone powers up.
- Configuration file retrieved via Trivial File Transfer Protocol (TFTP) at the same time as firmware updates.
- Choice of configuration file determined by a role defined in the corporate directory, often tied to job classification assigned by human resources at hiring or time of promotion.

- Checking of softphones to confirm that the PC or notebook is patched and has the latest approved versions of vLAN client, virus checker, and other applications.

As the size of a VoIP installation grows above 1000 the complexity calls for a management system that can deal with all phones at once (to push a patch) or groups of phones (enable a new feature for a user role or class of equipment, replace the configuration file on the call server). Best practice points to be policy-based managers that apply rules to determine which devices get to do what. Rather than configure each device, the administrator adjusts the policy. The NMS then updates configurations as required.

9.2 CONTINUOUS MONITORING AND MANAGEMENT

It is not enough to understand data path performance. Monitoring connectivity and watching the metrics associated with LAN data are necessary but not sufficient. Connectivity is a prerequisite to but no guarantee of good user experience on the telephone or video conference.

Better to ask key stake holders what they believe "success" is for them. Is it availability? Calls that connect quickly? Good sound or picture quality? Close synchronization between sound and picture in a video conference call?

What's important to the users, particularly the influential users, may determine which metrics to monitor and track.

- If application performance is important, the metric may be response time from the locations where users will consume the services. For example, time to enter a video conference or to start a video replay.
- If audio quality ranks high, the key network metrics regarding packet flows will be latency, jitter, and packet loss. These measurements can point to congestion points and misconfigured routers.
- Instrumentation located between WAN and LAN can look both ways to examine behavior of both nets.

Reporting works best if it can mark short bursts of errors or quality problems in real time, not just an average MOS for the day. Faults that generate an alarm and update to the dashboard must include QoS threshold alarms, high server or link utilization, latency, sound quality from a calculated MOS, and other key SLA metrics. Watching the MOS, for example, warns about a probable problem with user perceptions. The system should identify a cause, such as packet loss, for the MOS drop, then find the reason for the packet loss.

Trend reports on key metrics help isolate faults. They often predict problems before users report them. Plotted rolling averages over two years of statistics, or more, will help isolate temporary fluctuations.

Apply care when deploying appliances for monitoring the network and applications. Location within a domain is as important as test features and metrics. Collected data include:

- Statistics on utilization, traffic volume, applications or type of traffic, and the major sources and sinks of packets (the large users). These data support traffic analysis, historical trending, and planning.
- Sampled packets or entire packet streams for troubleshooting and forensics. With terabyte disk drives down to reasonable costs, all of an organization's traffic can be saved for weeks—every packet sent or received—for detailed forensic analysis.

Organizations subject to business regulations, privacy requirements, or security obligations have concerns that extend beyond telephony and UC, probably to all forms of information whether in transit or at rest. The impact on NMS can vary by industry, but planners will need to consider how to handle test recordings of packets. They almost certainly will contain personally identifiable information, health related information, or restricted data. As with live troubleshooting instruments that examine only the headers, an option is to truncate stored packets to remove their information content. The headers should provide the information essential to identifying faults yet occupy less disk space than full packets.

Be aware of the burden of NMS. Debugging modes built into routers, appliances, or software can help isolate problems. But these functions require CPU cycles, often a lot of them. Running an access control list on a router, with a filter to capture packets associated with an error form and a phone number, will sap 20 to 25% of the router's CPU cycles. You can turn off these features during normal operation, using them only for troubleshooting. Avoiding the impact on routers is a selling point for NMS appliances.

9.2.1 NMS Software

Many enterprises split the management job into pieces, based on the tools provided by each product vendor:

- Network infrastructure of switches, routers, and transmission lines.
- Servers and auxiliary equipment such as load balancers, intrusion detection/ prevention devices, and firewalls.
- VoIP and possibly other critical applications monitored for service quality (QoS) and not just connectivity or basic functioning.

Every major hardware and software vendor offers an "element management system" that deals with the configuration and monitoring of their product. Typically each runs on a separate server or management workstation. The better ones have a "northbound" interface, based on a standard protocol such as SOAP, that can feed into a higher-level manager of managers (MoM). Large computer vendors and multiple software vendors offer specialized MoMs. There are solutions that target the VoIP/UC environment specifically.

The MoM software tends to be complex, somewhat expensive, and requires considerable skill to discover the network topology, read the configurations of all manageable devices, and become aware of network services such as SIP trunks and leased lines. When set up, they offer a one-stop workstation to deal with the communications infrastructure, server hardware, and the applications on "one pane of glass."

As mention earlier, this leads to significant expense reduction. A market study by Nemertes in 2010 showed several significant results regarding the effect of having a MoM:

- The cost to identify a problem was reduced by more than 95%.
- Cost of staff time to resolve a problem dropped by more than 75%.
- Having more than one MoM was worse (a lot more expensive to find and resolve a fault) than having no MoM.

Voice is particularly difficult to manage because the quality experienced by end users depends on subtle interactions not only among the enterprise's servers (when not in a cloud environment) but also with the carrier service and many different end point devices. Unifying management allows the software to help find root causes.

9.2.2 Simple Network Management Protocol

Most communications equipment makers offer versions of their devices that allow remote management and configuration using SNMP. This is a very simple protocol, with commands to get or put values in the device's Management Information Base (MIB). Getting data is the same as reading a report, which can include anything that can be measured, but only what is in the MIB, such as bytes sent and received, errored packets, and port configurations.

Later versions of SNMP improved security significantly. Version 3 should be found in current devices. Older routers and switches, without updated firmware, may use Version 2. Version 1 has been obsolete too long to use any more.

Putting a value into the MIB, which requires the proper administrative rights, changes the configuration of that parameter. For example, a media gateway can change the gain in a voice channel or alter an IP address—again, almost any function or value. Some MIB parameters, statistics for example, are necessarily read-only. Each entry is a "managed object."

The IETF and IEEE define standard MIBs for devices such as routers, switches, and multiplexers. These are included in most network management (NMS) workstations and software. Hardware vendors are free to extend a MIB with proprietary features, such as a graphical representation of the device for display by the NMS, nonstandard features or functions, or values for parameters that fall outside the standard range. The extension exists as a file that the NMS imports and adds to the standard MIB.

Servers also have MIBs and make management information available for remote monitoring and configuration. Customarily configuration is done from an element manager or from the command line interface over telnet or remote shell connection.

Devices without SNMP are much less expensive than the functional equivalent with management. Even the smallest routers seem to be managed. In a small system, particularly at a single site, an unmanaged switch might be acceptable. The administrator is likely to work at that site and can handle a problem directly or just swap out the device. Any remote device should be manageable.

9.2.3   Web Interface

Many devices and most management workstations include a web server that allows a browser user to learn the status of a device or network. Many devices also accept input to modify configurations, adjust features, read logs, and so forth. This can be the best tool for password resets, or delegated to the end user.

In addition to admin rights to see and do everything, many call controller packages allow individual end users to control their own accounts and use the web interface to manage their phone. Users can see a call log for their phone, return missed calls or place new ones from the browser, create filters to handle incoming calls by source or time of day, and redirect calls to another phone based on time of day and day of week. The complete feature set is much larger and varies by vendor.

Web interfaces to the NMS may have to be on the open Internet, but if so should be protected with very strong authentication. An alternative is to restrict scope or authority of the Internet-accessible web server to read-only and require VPN or other secure access to exercise control.

9.2.4   Server Logging

Every server should run the system log file. This information has value in management analysis, offering insights into sources and sinks of traffic, peak usage periods, and so forth. But logs also have forensic value when troubleshooting network problems or examining security events.

Even when an intruder erases a log file to cover his activity, the absence of the file indicates a problem. Editing a log file to remove traces of an illegitimate activity takes time, so the presence of a log may discourage some hackers.

Maximum benefit of logs is hard to realize if they are read only by a human. Nobody can monitor a set of logs in real time. That's the job for a software tool designed to analyze system activity on the basis of log entries. Any event that is out of the ordinary triggers a message to the administrator, who can then focus on the problem. Constant monitoring should also detect attempts to modify or overwrite certain key files, possibly an indication of an intrusion.

An NMS workstation that can identify the time of a problem event greatly simplifies finding the associated log entries from their time stamps. Many networks set the clock in every device to GMT so there is no confusion about what time it is.

9.2.5 Software Maintenance

Nobody's perfect, and no software is either. There is always some way to improve an application or operating system and almost an endless supply of bugs and newly identified vulnerabilities. Updating all software is critically important, particularly on any device that could connect to the Internet. Scanning and probing from the Internet go on constantly, looking for vulnerable versions of software. You don't want them to find that your device has a vulnerability.

Software maintenance contracts offer automatic updates, which may work for smaller installations. Large enterprises often download new patches to a lab environment, where they are checked against the standard desktop and customer software applications. When approved, the updates are pushed to all the hardware platforms on the network from an internal server.

Testing injects a delay in deploying patches, which leaves open a possibility of attack, but can be warranted by a need to maintain business-critical applications. The trade-off balances the size of the opportunity for hackers against the risk of a patch problem halting important transactions (which does happen).

9.2.6 Quality of Service/Experience Monitoring

When an end user complains about poor voice quality or has a problem with image rendering, the root cause often is far from obvious. Sound defects may result from packet loss, server overload, high latency from router congestion, or other causes.

After confirming connectivity, application-level performance metrics can reveal some faults or point to other network levels. For applications like voice and video, where the perception of the user is the final metric, testing from the "far end" gives the user's viewpoint. It is possible for conditions to look fine from headquarters but not work well at the branch office.

Instrumentation devices, available as appliances or software for virtual machines, track performance at the application level, on top of connectivity. Appliances drop into remote sites, monitor traffic, then "phone home" with packet traces or statistics when queried. A good place to start is the point where the LAN meets the WAN. It's a concentration point that allows one monitor to see all the traffic at a site. Large sites or campuses may call for additional observation points inside the LAN.

NetFlow or a similar feature is available in routers, but the feature generally requires lots of CPU cycles from the router and can reduce a router's capacity (as measured by throughput or packets per second). The test feature can itself degrade quality; check by reading CPU utilization levels.

In addition to workstations in the Network Operations Center (NOC), software-based analysis tools are also available in "the cloud," as a service. Customers access the service from a browser to see who's doing what in what bandwidth.

Whatever the source of the information, a central workstation gathers statistics and logs events. The most sophisticated systems can go back in time to "recreate" an event, or at least allow analysis in considerable detail.

9.2.7   Validate Adjustments and Optimization

Many organizations maintain a test lab that matches the production environment in terms of server operating systems, router firmware images, and auxiliary functions such as DNS and firewalls. New software, firmware, and hardware gets tried in the lab before going into production.

Not as many organizations use the lab for trials of configuration changes but should, by conducting an acceptance test or at least a shortened version of it. Very large enterprises may extend this validation testing to any software or hardware product before it is approved for the standard desktop, notebook, or smart phone. That's good to do if you can.

9.3   TROUBLESHOOTING AND REPAIR

IP networks have best practices for finding and fixing faults. They will apply to one carrying VoIP. But additional methods provide insights into the special problems that voice can create, such as server overload, latency that's not noticed by data applications, delays in connecting calls due to firewalls and NAT, audible echo, one-way connections, and interactions with other servers (STUN, TURN, DNS, RADIUS, etc.). For these more complex problems, the troubleshooter needs some understanding of telephony, signaling, and applications as well as IP transport.

9.3.1   Methods

It never hurts to consider the old adage, "It's always the cable." Not true every time, but often enough to keep the saying alive. The cable from the user's terminal to the wall outlet is a good place to start. Visual inspection will reveal kinked or knotted wires or a silver-satin cable with RJ-45 connectors that found its way into 100 Mbit/s LANs. Older cable is suitable only for analog phones or low-speed serial data terminals, not Fast Ethernet. Structured cabling in the walls and data center are disturbed less so fail less often.

After the cable, look at the path. An initial evaluation should include an end-to-endconnection from the site of the reported problem to another phone drop. Measure in both directions as IP paths can be different each way. Look for packet loss, trace for asymmetric paths, check half/full duplex mismatch between adjacent Ethernet ports (most often at end points, possibly at internal points), and look for router queues not providing the proper class of service or packet fragmentation exceeding the MTU.

When testing, vendors recommend the technician starts from the end point where the problem report originated and work from the outside in. A problem shared by all the phones could reverse that order, to look first at the servers which could be causing the bad behavior.

Possible causes that can affect many users include:

- Phones don't activate when plugged in due to DNS server errors, bad configuration files, DHCP not configured properly to deliver default server addresses.
- Proxy server or SBC rewriting SDP content in SIP message incorrectly, for example, offering a codec that is not supported. See the discussion groups on the SIP Forum web site for reports of oddities among commercial products.
- DiffServe Code Point (DSCP) not set properly by end points, mishandled or not preserved by a router, or the values used are not consistent across domains or autonomous regions. Since this is a per-hop behavior, it is possible to locate the link with the problem.
- Route flapping in the network.
- Packet loss from congestion and queue misconfiguration. Some interactive games consume large amounts of bandwidth. People having too much fun are clues.
- Congestion from unexpected traffic, such as virus definition file downloads when a PC is turned on in the morning, killing VoIP during the transfer.
- Insufficient WAN bandwidth, particularly at branch office sites.
- Mismatch of half/full duplex setting on adjacent devices. Some NMS software can automate this test.
- Asymmetrical path routing; media channels behave more predictably if both directions lie on the same physical path. Traceroutes from both ends and some NMS consoles will show it.

Connectivity is only the beginning. After obtaining an IP address via DHCP, a newly installed IP phones picks up its configuration and firmware from a Trivial FTP (TFTP) server, which could be hosted with the call controller. The file must be named exactly right for the phone to retrieve it, and must contain accurate information for that phone at that location. Errors may prevent the phone from registering with the control server, registrar, or proxy. For example, the file may contain an inappropriate software image or incorrect IP addresses for key servers.

Sound quality complaints present a third level of difficulty and are harder to troubleshoot than connectivity. A good practice is to separate QoS issues from generic net problems. Path connectivity is visible in most network management systems (NMSs). But once that is established or confirmed, the problem may not disappear and will require additional, specialized expertise and probably different test equipment.

A person dedicated to "user experience" problems can be valuable in tracing issues related to voice and video quality. The search can lead to proxy servers on virtual machines, authentication servers for IP phones and users, or other aspects of the network and related infrastructure. For example, after-dial delay could be caused by a slow directory server.

Echo is unique. It may arise from electrical components, particularly gateways between VoIP and analog lines, or acoustic feedback from a phone or headset. It's not visible in packets or datascope traces, so the technician needs to listen to the conversation—sometimes considered a security vulnerability.

9.3.2   Software Tools

Sophisticated test equipment like protocol analyzers and datascopes can capture the full packets for voice, video, and facsimile for analysis. Most will play back audio or video and can display a fax page. Capture and playback of audio streams is necessary to troubleshoot echo. For other problems, or to avoid recording the message information, some testers can truncate each packet that is captured after the first 64 bytes, which collects the RTP/UDP/IP headers but not the payloads.

If an echo problem is repeatable, the out-to-in stepwise process on a test call will isolate the point that creates the echo. The cure may be more difficult, particularly if the cause is acoustic feedback. New hardware may be the best solution. Speaker phones and those designed for audio conferencing should not be a source of echo. They have built-in echo canceling, but it could fail or be turned off.

Any device that translates between 2-wire and 4-wire analog (such as an analog phone; POTS to microphone and earphone) contains a "hybrid" (Figure 2.2) that creates an impedance mismatch—a bump in the wire. Signals partly reflect off the "bump" to create the echo. A media gateway contains a similar function, interconnecting POTS and digital transmission. As a known source of echo, gateways include an echo canceler feature. The gateway should turn off echo cancellation (EC) when it detects a high-speed modem or facsimile machine on a channel to allow the modem to apply its inherent echo canceling. Failure to restore EC later on a voice connection will cause a problem in voice quality.

"Root cause" reporting by management software analyzes the error messages from equipment for comparison to a store of knowledge. Certain errors can trigger tests run by pre-written command scripts to further isolate the fault. You may see several possible causes, ranked by probability. The most sophisticated systems build up the knowledge base from experience on the actual network and rules added by network operators.

Always "test drive" after a change in configuration. You can automate larger-scope tests, more than spot checks, with a MoM or a script. Run a nightly check test. Proactive search for problems created by outside providers (change in routing path, modification in access rules, bandwidth reduction) can impact QoS for VoIP with no action by the enterprise.

Network and performance monitoring appliances normally impose no or low load (10 kbit/s each). They can, however, push a full load, such as for a simulated video transmission, on command. That test traffic can cause congestion-related symptoms if not turned off after the test.

Measure behavior for each administrative domain (autonomous region, AR) several times a year to establish a baseline or benchmark for performance and traffic volume. Gateways, session border controllers, routers, or an appliance can provide a good demarcation between ARs. Save those measurements to plot trends and for later analysis or comparison to real-time measurements when troubleshooting.

9.3.3 Test Instruments

Four-pair LAN cables need to be verified for longer reaches (nearing 100 m) or higher speeds (1 Gbit/s and above). A simple cable tester can reveal broken strands, shorts, opens, and mismatched pairs. Advanced testers qualify a cable by category (Cat 4, 5, 6, etc.) and/or bandwidth. Test with the final connectors in place as the quality of the attachment can limit overall quality of the cable.

A specialized test for the quality of electrical grounds will spot faults in what appears to the eye as an adequate connection. Corrosion, a hidden crack, or a loose fitting can raise the impedance of a ground to a level where it can't function as required. PBXs and other communications equipment should be grounded to a bus that is no more than about 8 ohms from true earth ground.

Optical cable requires its own form of tester. The Optical Time-Domain Reflectometer (OTDR) reports on dirty connections, tight radius bends, and misaligned fibers as well as a complete break in a strand.

Protocol analyzers come in the form of packaged instruments or as software to run on a notebook computer. While several important and commonly used analysis applications are open source and free, the speed and features of a formal test instrument easily justify their cost in appropriate situations.

Around the data center many tests require tools from the HVAC kit: thermometer, air flow meter, and pressure gauge. They are valuable when diagnosing overheating of equipment.

10

COST ANALYSIS AND PAYBACK CALCULATION

Vendors of IP PBXs point to rapid returns on investment at companies with at least five locations and 2000 people. Small organizations, under 100 phone extensions, may not see such a good return on investment:

- Hosted VoIP services charge "per seat" and typically include unlimited long-distance and local calling. The cost for that level of service can't be justified for a phone that used only for intercom calling on premises, but that may be the only option.

- The capital cost per seat for new phones, upgraded LAN, an IP PBX server, and additional network security measures can be daunting when spread across only a small number of users.

Smaller companies, particularly single-site businesses, show different economics. An online ROI calculator for SIP trunking (www.fliptosip.com) presents a selection of company sizes that starts with "5 to 3000" employees—a hint that very small companies may not be good candidates.

Startups with no previous investment are best situated for VoIP. They can build once for both voice and data (and video) services while ensuring sufficient capacity. There is no capital investment to write off nor any need to pay for removal of old equipment. And since new telephone equipment is almost all

VoIP and Unified Communications: Internet Telephony and the
Future Voice Network, First Edition. William A. Flanagan
© 2012 John Wiley & Sons, Inc. Published 2012 by John Wiley & Sons, Inc.

TABLE 10.1 Comparing capital costs of telephone systems

Expense	Scope	Count	Existing PBX	IP PBX (Owned)	Hosted VoIP	Hybrid	Other
Internet access setup	10 Mbit/s Ethernet		$	$	$	$	$
	T-1		$	$	$	$	$
	DSL/FO		$		$	$	$
	Other		$	$	$	$	$
Install voice trunks	PRI/T-1		$	$	$	$	$
	POTS		$	$	$	$	$
LAN upgrades	Routers		$	$	$	$	$
	POE switches		$	$	$	$	$
	Drops		$	$	$	$	$
Phones	Executive		$	$	$	$	$
	Manager		$	$	$	$	$
	Staff		$	$	$	$	$
	Lobby		$	$	$	$	$
	Softphone		$	$	$	$	$
VoIP control	Servers		$	$	$	$	$
Installation	Hardware		$	$	$	$	$
Software license	All		$	$	$	$	$
Structural buildout	Power, HVAC		$	$	$	$	$
Configuration	Software		$	$	$	$	$
Training	Staff		$	$	$	$	$
Contract cancellation fees	Existing services		$	$	$	$	$
Other	Leases		$	$	$	$	$
Totals			$	$	$	$	$

Note: Hybrid consists of SIP trunks and a gateway to a PBX or a VoIP system that reuses digital phones via specialized gateways between a VoIP call controller and digital phones from a previous installation of a legacy PBX.

TABLE 10.2 Comparing monthly recurring charges for telephone systems

Expense	Scope	Count	Existing PBX	IP PBX(Owned)	Hosted VoIP	Hybrid	Other
IP network access	10 Mbit/s Ethernet		$	$	$	$	$
	T-1		$	$	$	$	$
	DSL/FO		$	$	$	$	$
TDM trunks	PRI/T-1		$	$	$	$	$
	POTS[a]		$	$	$	$	$
Moves/adds	Phones		$	$	$	$	$
Management expense	VoIP service		$	$	$	$	$
Maintenance contract	Hardware		$	$	$	$	$
	Software		$	$	$	$	$
Leases	Hardware		$	$	$	$	$
			$	$	$	$	$
Internet access[b]			$	$	$	$	$
Costs which may vary	800 calls		$	$	$	$	$
	LD calls by travelers		$	$	$	$	$
	T&E for meetings		$	$	$	$	$
Totals			$	$	$	$	$

[a] Consider the need for POTS lines to meet availability targets or requirement for E911 compliance.
[b] Determine if the SIP trunk provider allows Internet access over the same local loop, charges an additional fee, or forbids it.

233

based on VoIP, this kind of firm's most attractive other choice could be a refurbished digital PBX.

Established firms with a PBX and digital phones are in a different position. While they likely pay for maintenance and moves/adds/changes on a monthly basis, that expense may be smaller than the cost of a new system. New features from UC and VoIP must justify a change.

With caveats expressed, it is still possible to reduce the cost of communications by switching from PBX's to VoIP and UC. Siemens fits the profile for the firm likely to benefit from VoIP and UC: thousands of employees, many locations with an installed PBX, and an international presence. Siemens reports impressive results from its own internal conversion of a global network that by 2010 replaced 130 PBXs with two data centers (on different continents) running call control servers. For 10,000 employees, costs dropped by $8 million the first year and $44 million per year after that—half the telecom budget.

Many vendors offer calculators to find your ROI from purchasing their solutions, hardware, or services. Because such tools intend to apply universally, they may not fit your situation very well. Examine the assumptions for the calculations from several angles. For examples:

- When comparing a PBX to a hosted service, is there consideration of upgrading the LAN to prioritize voice packets and deliver power over Ethernet?
- Does the calculation of server counts really provide sufficient capacity?
- Are software license fees and maintenance contracts specifically included? Are those licenses perpetual or per year?
- How do the bandwidth needs of the expected number of simultaneous users fit on the proposed access links at each site?

Tables 10.1 and 10.2 suggest items to consider when calculating the capital and recurring expenses associated with a telephone or UC system. Use them as a basis for comparing costs of various systems and vendors. Spend at least as much time on OpEx, the recurring charges, as you do on the capital costs. Nemertes Research found in talking to 1000 companies that capital costs are only 38% of the total in the first year. Over five years, CapEx is only 19% of the total cost of ownership. OpEx is the important cost.

Start a separate sheet for each set of assumptions: don't allow inconsistent numbers to appear together where they could be misunderstood when compared. If you anticipate a major change within a few years, try calculating that future incremental cost separately—it may not happen.

When calculating costs, look for potential events that could trigger large incremental costs. For example, if the original system is deployed near its maximum capacity, what are the costs associated with expanding to the next-larger software license or adding a cluster of servers?

Hidden savings may emerge as well. Do people on the road make long-distance calls on the phones in hotel rooms? What are travel and entertainment expenses associated with regular internal meetings held out of town? Local meetings impose a cost in time for participants to gather and return, particularly if driving in congested cities. Video conferences can reduce costs in all these areas.

To calculate the total cost of ownership (TCO), ongoing operating expenses are converted to net present value that assumes an interest rate on money (the internal rate of return). A firm's chief financial officer may specify a rate for use in TCO work at your company so that all projections are consistent.

License terms and costs may depend on more than one factor:

- Number of registered users, terminals, or end points (e.g., softphones, video cameras, and third-party devices).
- Capabilities of each terminal (Cisco assesses additional license "units" for some devices).
- Number of servers, which may count CPUs or cores as well as cabinets.
- Number of sites.

Be sure to include future license fees if the initial purchase is not a perpetual license that never needs renewal.

Every system requires network access for voice and data. VoIP allows these functions to share a single link, potentially reducing the number of links and costs. Alternatively, the same cost might allow for redundant links.

Telephone sets can be a major cost component of a new system. To minimize that cost, consider what minimum feature set will meet the needs of various groups, for example executives, managers, and staff. Reception areas, kitchens, loading docks, and similar areas may need only the simplest model.

Management each month incurs an expense whether the servers are owned and operated in-house, managed by a contractor, or outsourced entirely to a "cloud" provider. Is there a staff person able and available to manage servers and troubleshoot voice problems? If not, a contractor or managed service is more likely the better solution.

11

EXAMPLES OF HARDWARE AND SOFTWARE

This section intends to capture the state of the art in IP systems as this book goes to press. It is not an exhaustive catalog of available products, merely examples that represent capabilities and feature sets. Inclusion doesn't indicate an endorsement, and not appearing here means only that a product isn't listed here. When preparing to select a system, you may want to consult various comparisons of telephone systems like the *Sourcebook of Hosted and Cloud-Based VoIP and Unified Communications Services* at webtorials.com.

11.1 IP PHONES

IP phones from 10 years ago look inadequate today. Many early models were discontinued and replaced in a year or two because they lacked the CPU cycles, RAM, and interface controls to support all the emerging features buyers wanted.

The new models should last longer. They have many more buttons, larger displays, more powerful CPUs, enhanced software, and they conform to many of the standards and RFCs published in the last five years. The pace of developing new phone features in SIP continues steadily, but most of the essentials are now offered widely.

Few IP phone yet support IPv6, though that feature has appeared. With some planning, IPv4 phones will adapt to v6 with a software update. Because

VoIP and Unified Communications: Internet Telephony and the
Future Voice Network, First Edition. William A. Flanagan.
© 2012 John Wiley & Sons, Inc. Published 2012 by John Wiley & Sons, Inc.

their RAM or CPU may not be sufficient to run a double protocol stack of both IP versions, a phone operate only one at a time.

A primary differentiator of a phone is the number of line appearances it can handle. On the simplest phones that's one; "pro" desktop models can deal with 12. Attendant stations are either expanded with add-on busy lamp fields or go to the PC as a softphone with a flexible GUI for call management.

Phone operating systems come and go. At least four viable OSs currently support smartphones. Desktop phones are not as clear in marketing their OSs. Some employ "embedded software" that combines the OS and application into a unified firmware image.

Android phones and iPhones have thousands of optional applications available for inexpensive or free download. Many involve VoIP and mobile video conferencing. Smart phones with the appropriate radio hardware can participate in audio and video calls over 3G or 4G cellular, WiMAX, LTE, Wi-Fi, and Bluetooth (the short-range radio technology).

Most cell phones now have color displays to support web surfing, camera functions, and video chat. Desk sets have stuck with monochrome displays except on the high-end models, perhaps because they display mostly text. New introductions point to more color as a standard feature.

Unless stated otherwise, phones support one or both of G.711 and G.729 encoding. Some phones come with additional codecs. Later software versions may have different combinations of capabilities.

Will IP phones ever rival the longevity of the analog phone? Not likely, but they could match the digital desk sets that stayed in use for a decade or more. If your business processes don't change much, you may not have to replace all your phones in less than seven years.

Avaya pushes video as part of a UC installation. New products facilitate adding video, and the tablet-like Flare desktop end point runs a drag-and-drop user interface for setting up conference calls, the Flare Experience graphical user interface (Figure 11.1).

Avaya Flare with Flare Experience GUI   Low end Avaya

FIGURE 11.1   Avaya phones start simple (right) and go to the completely graphical tablet.

Even at the lowest end of the product line, **Avaya** phones have user-friendly features like back-lit display, four soft keys that adapt to the context, and navigation buttons.

Alcatel-Lucent has a conference phone that hosts up to four additional sites (Figure 11.2). It records sessions on a removable SD card. A-L stores four profiles, each with up to 1000 addresses and 20 preset call lists. SIP signaling, sRTP, power over Ethernet, and an internal web server—configure it from any browser.

Their's is a good example of a low-end phone. The display is 20 characters, any color as long as it's black. The 4008 model draws only Class 1 power from PoE. Other models draw more. The phone accepts a PKI certificate for sRTP.

In addition to SIP, A-Ls phones have the New Office Environment (NOE) protocol (Figure 11.3), used when registered with an OmniPCX IP PBX. The OmniPCX controls phone using either protocol.

FIGURE 11.2 Alcatel-Lucent conference room phone records onto an SD memory card.

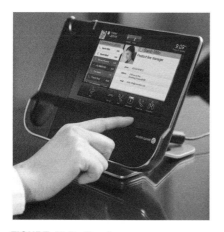

FIGURE 11.3 Touch screens are more common, like this A-L desk set (screen view at right).

FIGURE 11.4 Cisco supports multiple protocols, including a proprietary SIP version, Skinny.

FIGURE 11.5 Toshiba phones may be expanded with a busy lamp field module, right.

Cisco leveraged its position as sole vendor of network equipment to become one of the leading sellers of IP phones. Their most sophisticated models have color displays for multiple types of images and video, plus soft keys (Figure 11.4).

Toshiba offers a range of phones from a simple model with no display to one for the executives that has a large screen, more buttons, and a speakerphone. For attendants, or anyone handling calls for a group, there's a modern version of the busy lamp field that expands one of the phones (Figure 11.5). All the labels for the buttons are in the display—no more stickers.

11.2 GATEWAYS

While the PSTN exists, media gateways will necessarily bridge it to VoIP systems and networks. An enterprise may choose a large MGW for the main data center or a single-port MGW to attach a fax machine to a LAN. The variety is huge—more than 150 vendors.

FIGURE 11.6 Audio Codes media gateways can fit in branch offices and central offices.

With its only IP connection to the LAN inside the firewall, a MGW may not need to add IPv6 until the LAN converts.

Audio Codes MGWs range from tiny to carrier size (Figure 11.6).

Citel has specialized MGWs for keeping digital phones when changing from a circuit switched PBX to an IP PBX. Each box in a wiring closet terminates 12 or 24 existing drops on an RJ-21 connector (50-pin telco). It is compatible with digital phones by 8 of the major vendors and over 100 models, including digital, Centrex, and analog. The attraction is that there's no change in wiring (other than the patching to install the new boxes) and no training because the phones operate the same on a new SIP trunk as they did on a PBX. Each box has two FXO ports for local analog trunks to ensure proper handling of E911 calls.

DigiTalk converts signaling, broadly SIP to H.323. On the detail level, their SBC converts among the four methods of DTMF transmission:

- SIP INFO.
- IETF RFC 2833.
- In-band audio coding.
- H.323 Alphanumeric.

These devices make a case for an ability to route calls to multiple networks, minimizing cost and/or maximizing quality. While intended for carriers, a large enterprise can apply the same features for the same reasons.

11.3 SESSION BORDER CONTROLLERS

As an outward facing edge device, an SBC should be one of the first devices to run dual IP stacks, v4 and v6. At this time there is not much support for v6, but announcements indicate it will be widely available.

Acme Packet's SBC line (also sold by **Avaya**) announced IPv6 support in 2009, one of the first. This Net–Net Session Director product line focuses on IP–IP interfaces, including SIP trunking. TDM and circuit-switched services would be controlled by a different server, a gatekeeper or SIP proxy and a MGW (Figure 11.7). Security conscious government agencies and enterprises prefer the Net-Net 3820 and 4500 models which comply with Federal Information Processing Standard (FIPS) 140-2 for SIP-TLS, sRTP, IPsec, SSH, and SFTP without additional hardware. A Net-Net SBC can control session admission, bandwidth utilization, server loads, and SIP/H.323 session routing. Deployment options can reach high-availability for continuity of operations (COOP) or disaster recovery.

The **inGate** family of SIPerator session border controllers (Figure 11.8) ranges from branch office capacity (3 interfaces of 100 Mbit/s Ethernet) to carrier scale (6 GigEnet ports).

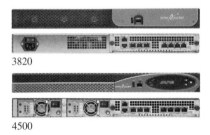

3820

4500

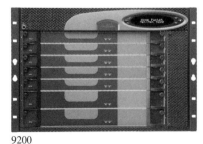

9200

FIGURE 11.7 Acme Packet leads the SBC market with full range of device capacities.

FIGURE 11.8 InGate has three packages for its session border controllers.

These SBCs combine many features that formerly would have run on separate devices:

- SIP proxy.
- SIP registrar.
- Network Address Translation (NAT/PAT).
- STUN server for far-end NAT traversal.
- Encryption, both TLS and sRTP.
- Authentication against an external RADIUS server.
- SIP trunking compliant with SIP Connect.

Cisco's Unified Border Element (CUBE) is a strong demarcation point between enterprise and carrier networks. Built on IOS firmware, it interoperates with Cisco's Unified Communications Managers and a SIP trunk. It also handles TDM services through MGW's and voice-enabled routers. As a software entity, CUBE may be added to certain installed Cisco hardware that accepts the Packet Voice DSP Module (PVDM2) for transcoding, conferencing, and voice gateway functions.

Sipera Networks has an SBC, but says that it's UC-Sec device is more, incorporating additional security features beyond the basics of the SBC (Figure 11.9). To secure SIP trunks the B2BUA arrangement is augmented by blacklisting, whitelisting, encryption of signaling and media streams, signature-based threat detection, and access control with policy enforcement plus their own two-factor credentials. That's on top of an application layer firewall, IDS/IPS, interworking between TLS and sRTP, and signaling translation to help SIP trunks work with a range of SIP proxies.

Essentially a software product, the SBC can run on may platforms to take advantage of hardware capabilities such as dedicated encryption chips, network processors, and high-speed interfaces.

VocalTec works at the high end, for interconnections between carriers that involve SS7 connections to a Service Control Point (SCP) that manages PSTN functions.

FIGURE 11.9 Sipera adds multiple security features to basis SBC functions.

GENBAND says it's S3 SBC is the first on the open, software-centric GENiUS ATCA platform. Advanced Telecom Computing Architecture standardized specifications for modular hardware (blade computers) optimized for CO equipment. GENBAND integrates call control, media sessions, signaling, security, applications, traffic management, and policy enforcement on a carrier scale intended to manage SLA's for multiple customers.

11.4 CALL-SWITCHING SERVERS

Every vendor of VoIP service necessarily operates a call server. Most VoIP carriers offer a desktop application that integrates with the switch and allows dialing, answering, filter setting, and other functions from the PC as well as the phone. This application is often called a "call controller" or "call manager." They are not the switch, proxy, or gatekeeper that is the subject of this section, many of which are called communications controller or call manager.

The user seldom gets a choice of PC app; it comes with the service. It will not be covered here.

Enterprises may choose to run in-house the same or similar switch software used by carriers. The software may be hosted on a rented server in a third-party data center. To outsource all the management responsibilities, let a hosted service take care of servers, operating systems, IP PBX software, and network access—so only the phones remain on premises, connected to the LAN.

If running your own servers, you should consider what technical talent will be needed. Not only is there a quantity issue (more servers = more staff) but there are the qualities needed to manage voice servers and possibly multiple operating systems. Many products related to VoIP and UC, not just those from Microsoft, run on a Windows operating system. Most of the open source applications and many of the high-availability products run on Linux, with Windows also supported by some but not all software vendors.

Servers exposed to the Internet to provide VoIP services, such as a proxy, TURN, or STUN server, should be hardened and maintained to remove vulnerabilities when discovered. Hardening is outside of scope, other than to summarize the major steps:

- Remove unneeded applications, daemons, and processes.
- Virtualize the server, with a virtual machine on the same hardware running intrusion protection.
- A good practice lately is to add a whitelisting module that prevents unknown applications (such as new malware) from running.

Choice of a deployment option should consider availability, or expected uptime. IP phones can be managed by open source software on a cheap PC. That arrangement has many failure points, any one of which will stop phone service. The prudent choice involves at least two servers (not PCs), each of which may have multiple CPUs, disks, power supplies, and network interfaces.

Redundancy is a good start, but an automatic recovery mechanism is better. Configuration of redundant components can enhance reliability. For examples:

- Create a cluster of virtual machines (VM) for each application or function, based on a hypervisor (software that supports multiple VMs on one hardware set).
- Put at least one VM on each physical server.
- Assign more than one CPU core (or a portion of its capacity) to each VM.
- Set up multiple disks as hot-swappable RAID arrays rather than a single file system.
- Install redundant power supplies in each chassis; put at least one on a UPS.

The application software will also affect how the system recovers from a failure.

VoIP vendors recognize the importance of availability and almost every one offers a feature called something like automatic recovery. There are major differences, which may impact a choice.

At the top end, fully redundant hardware platforms share the load and multiple software instances maintain knowledge of every session. That is, each machine has all the state information for all users. Upon a hardware failure or a network outage to one of the servers (which, ideally, are located in separate data centers) the application continues to run without interruption or loss of function for any user. Nobody knows there's a failure without hearing an alarm from the management system. Virtualization is necessary to make this method work.

Ideally, statefull redundancy is the goal for telephone service. An enterprise can set it up, hire a contractor to do it, or outsource the task to cloud provider.

Hot standby doesn't require quite as much complexity in configuring OSs as there may be no load sharing. The backup server may or may not learn the state of users and sessions. The fail-over happens shortly after keep-alive messages stop coming from the failed machine—perhaps in 30 seconds. The standby then starts handling requests. There may be no service for that half minute, but calls in progress continue if they don't request anything from the server. If the BU keeps state data, all calls can continue normally, using additional services after the fail-over completes. If the BU didn't have call state, and during fail-over, any change request will return an error from the server (call ID not known, or similar), causing the end points to drop the session.

Early digital PBXs had a few more features than the analog switches they replaced. The chance to gain competitive advantage led to a feature race, creating more than 500 named features. Some were obscure: executive override of no−barge-in, for example. When vendors of circuit switches migrated to VoIP, they brought the software and almost all 500 features. Standard SIP functionality is still catching up. Dozens of draft projects are developing extensions that will add old PBX features and quite a few new ones.

Examine the feature list carefully to ensure the system under consideration support those features you need. Get a demonstration to confirm each function. Get the demo on the software version to be delivered and not demoware.

11.4.1 IP PBX

Starting alphabetically or from lowest cost, the first software product would have to be **Asterisk** (www.asterisk.org). This open source IP PBX software package is championed by **Digium**, a manufacturer of PC cards that make a PC into a media gateway as well as a call controller. When running on a Linux operating system, Asterisk can take advantage of the clustering features included in major operating systems, spreading the load across multiple servers with automatic failover—a way to achieve the high availability expected from telephones.

Increases in CPU speed allow consolidation of multiple server functions into a single device. **Alcatel's** BiCS communications controller for the SMB market puts all functions into a pizza-box package (Figure 11.10).

It performs call control, automatic call distribution (ACD), fax handling, chat, presence, and messaging for up to 1000 users. Alcatel used DECT phone technology for on-premises mobility.

When **Avaya**, one of the traditional circuit switch companies, transitioned to VoIP, it brought extensive telephony experience and a large feature set from its legacy PBX line (including voice and data products acquired from Nortel's bankruptcy). Digital PBXs run on software control, so they had an advantage in owning code that performs all PBX functions. They migrated some features to follow standards, but continue to expand proprietary functionality. For

FIGURE 11.10 Alcatel-Lucent, like many vendors, offers a rack-mount server for call control.

example, in 2011 they introduced a wireless tablet that controls all the functions of the desk phone and delivers video conferencing, documents, and still photos to a graphical user interface. From its traditions, Avaya developed a call control solution that grows to very large size and is highly available with only two or three servers. The architecture is similar to a central office switch.

Avaya presents its Aura Collaboration Server as an easy step into video conferencing. ACS runs on a separate server that joins a LAN and uses SIP to participate in sessions. Through server virtualization, the ACS may share a server with the Aura.

BroadSoft's call control software has been the foundation for multiple VoIP service providers. Very large enterprises also use it. A web portal provides administrative control but also gives each end user a panel to monitor calls, place calls, configure filters for handling incoming requests, and setting up forking to multiple phones. As a carrier product, the server partitions into administrative domains which could be ISPs, corporate divisions, or regional offices.

Cisco has a major market share for its VoIP solution. This brand often is the default choice in single-vendor shops. While it is primarily an IP networking company, in the decade since it got into telephony Cisco acquired companies for their expertise and technology related to voice, and advanced its product functionality considerably. The products have grown in capacity by incorporating more servers in clusters, and networks of clusters. The largest possible system requires many separate servers.

Siemens Enterprise Communications, another descendant from a circuit switch company, now offers a software foundation for UC: OpenScape UC Server. It runs on any commercial hardware, which means that with Resilient Telco Platform middleware, virtual server clustering, and active-active dual data centers the availability can be pushed as high as an enterprise needs. The CO legacy lives on in that high availability feature but also in scalability to 100,000 users on as little as one pair of redundant Linux servers. A compact PCI chassis is an option.

Siemens EC published list pricing (2011) for user licenses. The complete enterprise software (up to 100,000 users) including OpenScape Voice and OS

Applications (the UC part) is licensed as a bundle of all features with a list price of €165, not including hardware.

For the smaller to mid-size versions (up to 150, 350, 500, or 1000 users) OpenScape Office is offered as an appliance as well as software for standard or virtualized servers. With all the feature applications integrated into a standard server, the appliance installs quickly. An analog media gateway control function handles facsimile machines, analog phones, and legacy TDM trunks (ISDN, T-1) via MGWs. A bundled user license for all feature options and the hardware lists at €87. The seat license for the 350-user appliance is €137.

ShoreTel takes a modular and distributed approach to call switching. Each unit (Figure 11.11) supports a number of end points—IP phones, analog phones, trunks—in a half-rack width 1U box. When managed from the web-based ShoreTel Director, all the hardware appears as a single system. In release 12 of its software **ShoreTel** added High Definition (HD) audio conferencing, zero-download desktop sharing, XMPP-based Instant Messaging, presence, Microsoft Outlook scheduling, and multimedia recording to existing capabilities for audio and web conferencing. They also doubled the maximum number of users to 20,000 on this distributed system.

Toshiba software runs on standard servers. The company offers two form factors, desktop (on the shelf) and rack-mounted (Figure 11.12).

11.4.2 Conference Bridges/Controllers

Sonexis offers a heavy (65 lb) and heavy duty ConferenceManager for audio and web meetings in a 2 RU package. It scales from 12 to 600 ports of audio

T-1 trunk + MOH in + paging out

50 IP phones + 4 FXS + 2 FXO

FIGURE 11.11 ShoreTel's distributed architecture adds half-rack width boxes to expand capacity.

FIGURE 11.12 Toshiba platforms are rack mounted or packaged for a desktop or shelf.

and 5 to 600 seats of web. Simplicity of use is emphasized, starting with installation in under an hour in most environments—VoIP (SIP or H.323) or PBX (CAS or ISDN). A key feature is control of attendees so that confidential discussions can be held in secure confidence.

The feature list is worth examining for ideas that can improve business processes:

Audio Conferencing

- 12 to 576 ports PSTN (PRI, CAS).
- 12 to 600 ports VoIP (SIP).
- 12 to 200 ports VoIP (H.323).
- Web-based host control panel.
- Real-time speaker identification.
- Integrates with Outlook and Lotus Notes.
- Ad hoc (reservation-less) or scheduled conferences.
- Automatic email invitations.
- IVR for greetings and help menu.
- Multiple conference announcement options (tone, record name, none).
- Audio recording.
- Attendees connect via dial-in or dial-out.
- Emergency conferencing and blast dial-out.
- Supports up to 10 subconferences off the main conference.
- Generates call detail records.

Web Conferencing

- 5 to 600 seats.
- Shares applications and documents.

- Cross platform and no download for participation.
- Interactive Whiteboard.
- Annotation tools.
- Follow-me web browsing.
- Notepad.
- Text chat.
- Q&A, polling, voting.
- Quick invites.
- Authenticated host login.
- Saved conference rooms.
- Full-screen presentation.
- Integrated with audio bridge manager.
- Ad hoc (reservation-less) and scheduled.

11.4.3 Call Recorder

Toshiba clearly states the functionality of a call recorder in its description:

- Cradle to Grave Recording captures every moment of the call, even while the caller is on hold.
- Look-Back Recording records an entire conversation, even if the recording was initiated after the call began.
- Call Monitoring allows authorized users to monitor calls as they're happening, whether those calls are being recorded or not.
- Selective Recording records specific or random extensions, groups of extensions, or every extension.
- Bookmarks are automatically inserted when calls are transferred or put on hold. Particular calls or parts of calls can be easily and quickly found.
- After-Call Actions can be taken after a call, including sending an email or instant text message, or launching another program.

In various systems the actual recording of packets to disk may be in a separate device. The control software duplicates packets and directs the copied stream to the recorder.

The **Oaisys** software, Talkument, is called "voice documentation software." It works as a collaboration tool as well as an archive for phone conversations. Recording can be turned on for only selected extension (VoIP or TDM). An API integrates with major telephone systems and messaging applications so recorded calls may be sent as attachments to emails very easily. Their Tracer application records screen images with the voice content. For use as evidence and to help correlate recordings with other events, the play back can include an audible spoken timestamp.

11.5 HOSTED VoIP/UC SERVICE

Organizations that don't want to buy or manage their telephone or UC system will find many vendors offering to host in their own data centers all the proxy servers, registrars, media gateways, and authentication services their customers need. They may also provide an access circuit to their data center, security tools and appliances, eNum directories, and other support.

This market changes so quickly that there is no point in comparing vendors or prices in a book. The forms provided at the back will help in collecting the information needed to make a comparison in your situation at the time you need to decide.

Many services share features because they use the same software, often from Broadsoft. Branding likely hides the platform.

You should expect at least the following:

- SIP trunking from the public Internet (you provide the access circuit) or on a dedicated access circuit.
- Incoming calls route to a specific IP phone.
- Forking of incoming calls to at least three phones.
- Redirection of incoming VoIP calls to PSTN or legacy phone: ring/no answer, busy, or always. Some services don't offer traditional features, such as trunk hunt groups, but a combination of forwarding and simultaneous ringing may provide a workaround.
- Call history showing in- and out-bound calls, missed calls.
- Voice mail; pick up as audio by phone or from browser, audio file attached to email, or text message (voice-to-text conversion). Beware of short time limits on message length.
- Call filtering by calling phone number, called phone number, time of day, day of week
- User configurable preferences and filters

Hosted PBX pricing can be hard to compare to POTS lines. Service providers offer "pricing plans" with numbers of calling minutes to defined geographical areas (United States, Canada, Puerto Rico, in one case). But that monthly fee is far from the total cost.

If signing up on line, be sure you select the plan you intended. Some sites are laid out to encourage taking the more expensive offerings.

You may not find an "all-in" quotation on a website, or it may be hard to locate, or it may be avoided entirely. The amount of the FCC line charge wasn't found on a couple of checked sites, not included in these examples.

Mitel AnyWare service is a virtual PBX that costs a fixed fee per month per user, minimum of ten users. The starter rate ($25 at this time) includes local calling, no LD. An additional $10/month buys unlimited LD. Dedicated fax is

$4/month per line. A mobile feature brings cell phones into the system with the same features as the hard wired desktop phones.

Fonality aims for the business market with unlimited calling plans. Residential plans include hundreds of minutes, in various tiers.

Microsoft's purchase of **Skype** may open more doors into business users for the no-charge peer-to-peer phone and video service. It offers international and off-net calling for a fee.

11.6 MANAGEMENT SYSTEMS/WORKSTATIONS

The problem is that the complete solutions need to climb a long learning curve. The specialized managers of a single product start quickly but don't provide the breadth of monitoring and control that a large deployment needs.

Management tools abound, offered by the biggest vendors and the newest startups. The known names in the industry are easy to find. The new products may not be mature enough. There are still many in the middle; these are examples.

Clarus Systems' management software, ClarusIPC Plus+, governs a VoIP installation end to end. The company says it is useful during initial deployment and upgrades as well as for ongoing monitoring, testing, and reporting. It's optimized for Cisco equipment.

Packet Design stresses route analytics, the real-time evaluation of network paths at the router level (IP or L3). By listening to router update messages, the appliance follows paths through multiple domains, including the LAN and MPLS carrier networks. It builds a model of the entire network, similar to the understanding routers have about network topology. Great troubleshooting tool.

Tone Software's ReliaTel management suite is attuned to the special needs of VoIP and UC as these technologies take over from legacy services:

- Manages the entire communications environment: TDM/SIP and trunks/devices.
- Promotes VoIP sound quality through real-time analytics and diagnostics.
- Monitors conformance with Service Level Agreements (SLAs) on Network Health Dashboards.
- Holds a Knowledge Base of alarms and their resolutions to speed fault isolation.
- Anticipates service issues by reporting trends in capacity utilization and traffic.

12

APPENDIXES

12.1 ACRONYMS AND DEFINITIONS

Some terms are defined at the locations referenced in the Index.

0x	Indication that the following characters are a hexadecimal expression.
5ESS	Class 5 Electronic Switching System, CO voice switch that serves end users.
AB	Two signaling bits in T-1 CAS when using a superframe.
ABCD	Four signaling bits in T-1 CAS when using the extended superframe.
ABNF	Augmented Backus–Naur Form.
AC	Area Code, in telephone number.
ACD	Automatic Call Distributor, PBX function or machine to spread calls among phones.
ACELP	Adaptive CELP.
ACF	Admission Confirm.
ACK	ACKnowledge, message type to confirm receipt of request or action taken.

VoIP and Unified Communications: Internet Telephony and the
Future Voice Network, First Edition. William A. Flanagan.
© 2012 John Wiley & Sons, Inc. Published 2012 by John Wiley & Sons, Inc.

ADC	Analog-to-Digital Converter, first step to encode voice as a digital stream.
address	IP address of a host; MAC, DLCI, or other L2 destination or source identifier; cf socket.
ADPCM	Adaptive Differential Pulse Code Modulation, form of compressed voice encoding.
AH	Authentication Header, form of IPsec that digitally signs but doesn't encrypt.
A-Law	Form of PCM companding used with E-1 links.
ALF	Application Level Framing.
ALG	Application Layer Gateway.
ALI	Automatic Location Information (Identifier), geographic indication associated with a telephone, landline or cellular, delivered to a PSAP with E911 call; kept in an ALI database.
AN	Acknowledgment Number, associated with an octet in a TCP stream or a packet.
analog	Can take on any value to represent sound; *see* digital.
ANI	Automatic Number Indication, caller ID, CLID.
ANSI	American National Standards Institute.
AOR	Address Of Record, URI that points to a domain with a DNS service; "public address" of the user.
API	Application Programming Interface, set of commands and procedures between software entities.
AR	Autonomous (Administrative) Region, portion of a network under one administrator.
ARJ	Access Reject.
ARP	Address Resolution Protocol, how L2 devices associate IP addresses with MAC addresses.
ARPA	Advanced Research Projects Agency.
ARQ	Admission Request.
AS	Autonomous Systems, network recognized in BGP routing tables; number assigned to an organization by IANA.
ASCII	American Standard Code for Information Interchange, defines 7-bit code (usually plus parity) for printable letters and symbols plus control characters.
ASCONF	Address Configuration Change Chunk, in SCTP.
ATCA	Advanced Telecom Computing Architecture, standardized specification for modular hardware optimized for CO equipment.
ATIS	Alliance for Telecommunications Industry Solutions.
ATM	Asynchronous Transfer Mode, cell relay, transmission in 53-byte cells.
A/V	Audio/Visual.
AVP	Audio-Visual Profile, SIP configuration template.
AWG	American Wire Gauge, approximately equal to 1 inch divided by wire size.

B2BUA	Back-to-back User Agents, a form of proxy server or SBC.
B8ZS	Binary 8-Zero Suppression, substitutes 000+-0-+ for 00000000 to maintain ones density on T-1 line.
B Channel	Bearer channel in ISDN for payload, voice, video, and so forth, but not signaling.
BGP	Border Gateway Protocol, finds paths between autonomous systems.
bit/s	bits per second, unit of digital transmission speed.
BNF	Backus–Naur Form, symbolic method to define elements and express syntax.
botnet	A large number of PCs (a network) infected with malware that turns them into robots (bots) under the control of a botmaster.
BRI	Basic Rate Interface, 2 bearer+1 data (signaling channel) on local loop.
b/s	bits per second, also bit/s.
B/s	Bytes (octets) per second.
BTN	Billing Telephone Number, the main DN of an account.
b/w	Bandwidth.
byte	Octet, 8 digital bits.
CA	Certificate Authority, server that creates, signs, and verifies PKI certificates for other servers and end users.
CAC	Call Admission Control.
CALEA	Carrier Assistance for Law Enforcement Act.
CAMA	Centralized Automatic Message Accounting, billing system; ISDN-like trunk between PSTN switch and billing system also used to deliver caller ID into PSAPs.
CapEx	Capital Expenditure, costs that have to be depreciated rather than expensed.
CAS	Channel-Associated Signaling, robbed-bit signaling on a channelized T-1 line.
CB	Channel Bank, digital multiplexer for 24 audio channels.
CBN	Call Back Number, related to E911 call, for PSAP agent to use.
CELP	Code-Excited Linear Prediction, form of compressed voice encoding.
CERT	DNS record containing a PKI certificate.
CID	Conference ID.
CLID	Calling Line IDentification (also CLI).
client	Issues commands in SIP.
CM	Configuration Management.
CN	Comfort Noise, sound from telephone handset to confirm line is not dead.
CNAME	Canonical NAME, identifies a source in RTP such as a host, which may apply to multiple streams or sessions.

context   An association of terminations on an MGC (H.248.1 §6.1).
CO   Central Office, where the carrier keeps its switch and transmission equipment; the upstream end of the local loop.
Core   Set of essential functions.
core   One of possibly multiple microprocessors on a single CPU chip.
COTS   Commercial Off-The-Shelf, software or equipment that's not customized.
CPE   Customer Premises Equipment.
CPU   Central Processor Unit, the calculating engine in a computer.
CRC   Cyclic Redundancy Check, an error detection scheme, for ARQ or frame/cell discard.
CRLF   Carriage Return and Line Feed, two ASCII characters that create a blank line.
CRV   Call Reference Values.
CSeq   Command Sequence number, in SIP commands.
CSMA   Carrier Sense Multiple Access, when with Collision Detection it's Ethernet.
CSRC   Contributing SouRCe, secondary input(s) to composite stream (RTP).
CWR   Congestion Window Reduced.

DAC   Digial-to-Analog Converter, decodes voice from digital to sound.
D Channel   Data channel for signaling in ISDN.
DB   DataBase.
DB-25   Connector for RS-232 serial electrical data interface.
DCF   Disengage Confirm.
DDDS   Dynamic Delegation Discovery System, part of eNum.
DECT   Digital European Cordless Telecommunications, wireless TDM phone service.
DHCP   Dynamic Host Configuration Protocol.
Dialog   Persistent relationship between two end devices; identified by the Call-ID value, a local tag, and a remote tag.
DID   Direct Inward Dial.
digital   Representation of sound or image limited to bits of value 0 or 1.
DISA   Direct Inward System Access.
DLCI   Data Link Connection Identifier, link address in Frame Relay header.
DN   Directory Number, network address used to reach called party (POTS, ISDN).
DNI   Dialed Number Indication, the extension information delivered with a DID call.

DNIS	Dialed Number Indication Service, what you pay for to get DNI.
DNS	Domain Name Service, converts URL to IP address for each specific service.
DNSsec	DNS SECurity, extensions to the DNS protocols that allow servers to authenticate each other.
DoS	Denial of Service, form of Internet attack.
DPNSS	Digital Private Network Signaling System, PBX interface for common channel signaling; used in UK.
drop	Cable that reaches the customer; from a pole or the wall.
DRQ	Disengage Request.
DS-0	Digital Signal level Zero, 64 kbit/s TDM channel.
DS-1	Digital Signal level 1, 1.544 Mbit/s, the T-1 rate.
DS-3	Digital Signal level 3, 28 T-1's multiplexed on one line (672 channels).
DSCP	Differentiated Services Code Point, value of the TOS field in IPv4 header that indicates a class of service requested for that packet (RFC-2474).
DSL	Digital Subscriber Link, local loop transmission on one UTP.
DSP	Digital Signal Processors.
DTLS	Datagram Transport Layer Security, encrypts SCTP sessions.
DTMF	Dual-Tone Multi-Frequency, TouchTone signaling.
E-1	European DS-1 of 32 DS-0's, uses one each for framing and signaling.
E.164	ITU Recommendation, how PSTN phone numbers are distributed.
EA	Enterprise Architecture.
EC	Echo Cancellation (Canceler).
ECNE	Explicit Congestion Notification Echo.
ECRIT	Emergency Context Resolution with Internet Technology.
EIA	Electronics Industries Association.
ELIN	Emergency Location Identification Number.
E&M	Ear & Mouth, signal leads on a 2- or 4-wire analog voice interface, usually a trunk.
Enet	Ethernet.
ENUM	Electronic NUmbers Mapping, representation of E.164 phone number as FQDN.
EOL	End Of Line, 12-bit symbol in fax message to separate scan lines (T.4).
ERL	Emergency Response Location.
escape	In a character stream, a reserved character that indicates a change in meaning for the "escaped characters" that follow immediately.

ESF	Extended SuperFrame.
ESGW	Emergency Services Gateway.
ESP	Encapsulating Security Payload, encrypting form of IPsec.
ETSI	European Telecommunications Standards Institute.
EV-DO	EVolutionary, Data-Only (Optimized); form of broadband mobile access service based on CDMA cellular technology.
F	Frameing bit on a T-1 interface.
FCC	Federal Communication Commission.
FEC	Forward Error Correction.
FIN	FINal, 1-bit indicator field in TCP header of packet that closes connection.
FIPS	Federal Information Processing Standard, various standards cover security measures, governance practices, etc., for networks and data centers.
FoIP	Fax over IP.
FTTH	(Optical) Fiber To The Home.
FQDN	Fully Qualified Domain Name, which includes the top level domain after the last ".".
FR	Frame Relay.
FRF.n	Frame Relay Forum implementation agreement, number n.
frame	Layout of bits in digital transmission that allows receiver to identify channels.
FTP	File Transfer Protocol.
FXO	Foreign Exchange Office.
FXS	Foreign Exchange Subscriber.
G3	Group 3, most common type of facsimile machine.
generic	In a PBX, a specific version of the base software.
GigEnet	gigabit/second Ethernet.
GK	GateKeeper, controller in H.323.
GCF	Gatekeeper ConFirmation.
GPS	Global Positioning System, delivers precise time as well as locations.
GRJ	Gatekeeper ReJection.
GRQ	Gatekeeper ReQuest, message from end point to register with H.323 GK.
GSM	Group Speciale Mobile (Global System/Standard for Mobile Communications), a 2G CEPT (ITU) standard on digital cellular.
GUI	Graphical User Interface.
GWC	GateWay Controller, manages the MGW.
HA	High Availability, uptime/(total clock time) near 100%.
HDLC	High-level Data Link Control, frame format and procedure originally for terminals on multidrop lines.

HDvoice   Encoding that captures wider frequency range than PCM (e.g., 300–7000 Hz).

header   Lead portion of a protocol frame that contain addresses and control information; statement within a signaling message to convey the value of a parameter.

hop   Travel from one device to another, as between routers; the up and down legs of a satellite circuit.

HR   Human Resources, the personnel department.

HTML   HyperText Markup Language.

HTTP   HyperText Transmission Protocol.

HVAC   Heating Ventilating and Air Conditioning.

Hz   Hertz, the unit of cycles per second; applies to all frequencies.

IAF   Internet Aware Facsimile, machine with an IP interface, and not necessarily a modem.

IANA   Internet Assigned Numbers Authority.

ICANN   Internet Corporation for Assigned Names and Numbers.

ICE   Interactive Connectivity Establishment.

ID   IDentifier, identification.

idempotent   Takes on one value; occurs only once.

IDS   Intrusion Detection System.

IEC   International Electrotechnical Commission, standards body.

IEEE   Institute of Electrical and Electronic Engineers.

IFP   Internet Facsimile Protocol.

IGP   Interior Gateway Protocol, routing protocol used within an administrative domain or autonomous system.

IM   Instant Messaging.

IMS   IP (Internet) Multimedia S(ubs)ystem, hardware and software to allow a carrier to deliver voice, video, and data over a data infrastructure.

INV   INVite, message in SIP to set up a new call or add to conference.

IP PBX   Software, possibly integrated on a server or custom hardware, that functions as a SIP proxy and registrar, or H.323 gatekeeper, to route calls for IP phones.

IPHC   IP Header Compression.

IPS   Intrusion Prevention System, detects but also acts against bad behavior.

IPsec   IP security, extensions to IP for digital signing and encryption.

IPv4   Internet Protocol version 4, the dominant form on the Internet; 32-bit addresses.

IPv6   IP version 6, increases each address field to 128 bits.

ISDN   Integrated Services Digital Network, switched TDM and packet services.

IS-IS	Intermediate System to Intermediate System, an "interior" IP link-state routing protocol.
ISO	International Standards Organization.
ISP	Internet Service Provider.
ISUP	ISdn User Part, signaling layer and protocol for call control (SS7).
ITSP	Internet Telephone Service Provider.
ITU	International Telecommunications Union.
IVR	Interactive Voice Response, machine that asks you to talk to it.
IXC	iItereXchange Carrier.
JID	Jabber ID, address for instant messaging bare: <localpart@domainpart> full: <localpart@domainpart/resourcepart>
k	1000, prefix indicating multiplication by 1000; applied to bit rates, line speeds.
K	1024, even power of 2, applied to memory size, etc.
km	kilometer.
kbit/s	kilobit per second, 1000 bits per second.
KPML	Key Press Markup Language, method to encode DTMF signaling in VoIP.
LAN	Local Area Network.
layer	A protocol within a stack of protocols such as the ISO model.
lD	Long Distance, a connection that crosses between carrier areas.
LDAP	Light-weight Directory Access Protocol.
LDP	Label Distribution Protocol, one way to configure path in MPLS network.
LEC	Local Exchange Carrier, provides phone service and assigns E.164 phone numbers.
LIS	Location Information Server.
LNP	Local Number Portability.
local loop	Connection between customer site and carrier's central office; a trunk.
lr	Loose routing, indicator in some SIP headers.
LS	Loop Start, form of trunk signaling on a POTS line.
LSD	Least Significant Digit.
LSP	Label Switched Path, specific route in an MPLS network.
LTE	Long-Term Evolution, pure-IP 4G mobile broadband standard.

MAC	Media Access and Control, functions of the Ethernet headers.
MAC	Moves Adds and Changes, largest maintenance work effort for PBXs.
mark	logical representation of a digital 1 bit.
Mbit/s	megabit per second, 1 million bits per second.
MCU	Multipoint Control Unit, conference bridge in H.323 network.
Megaco	MEdia GAteway COntrol protocol.
MF	Multi-Frequency, signaling based on tones in the voice channel.
MGC	Media Gateway Controller.
MGW	Media GateWay.
MIB	Management Information Base, DB of configurations and statistics about a device.
MID	Message Identifier.
MIME	Multipurpose Internet Mail Extensions, various methods to encode files for transmission on the Internet.
MOH	Music On Hold.
MoM	Manager of Managers, master workstation over element managers.
MOS	Mean Opinion Score, quality rating for voice transmissions $(1-5)$.
MPEG	Motion Picture Experts Group, defined digital encoding for movies/videos.
MPLS	MultiProtocol Label Switching.
MRC	Monthly Recurring Charge.
ms	millisecond, 1/1000 of a second.
MSAG	Master Street Address Guide.
MSD	Most Significant Digit.
us	microsecond; 1/1,000,000 second; also writen with mu: μs
MSRP	Message Session Relay Protocol.
MTU	Maximum Transmission Unit, largest packet size allowed.
mu-Law	Form of PCM companding used with T-1.
mux	Multiplexer, combines multiple streams to share a link.
MWI	Message Waiting Indication.
MX	DNS record that contains the address of an email server.
NAT	Network Address Translation.
NAPT	Network Address and Port Translation.
NAPTR	Naming Authority Pointer Record, a DNS response.
NIC	Network Interface Card, these days usually Ethernet on UTP but can be optical.
NMS	Network Management System.

NOC   Network Operations Center, help desk and engineering support.

noise   Component of audio or electrical signal other than wanted information.

nonce   Number used ONCE, something added before encrypting or for another purpose; a generated number to identify a transaction.

NS   Name Server, the authoritative DNS server for a zone.

NSDU   Network Service Data Unit.

NSInet   Emergency services IP network for E911 and first responders.

NTP   Network Time Protocol, transfers wall clock time from master server to other hosts.

OC-3   Optical Carrier level 3, SONET rate of 155.52 Mbit/s, matches STS-3.

octet   8 bits; a byte.

OpEx   Operating Expense, recurring costs of a project (see CapEx).

OS   Operating System on a computer or smart phone.

OSPF   Open Shortest Path First, a link-state routing protocol that accounts for link bandwidth, congestion, etc.

OTDR   Optical Time-Domain Reflectometer (Reflectometry), instrument (method) that locates faults in optical fiber from light reflected back at the tester.

packet   Group of bits or digital characters transmitted together as a unit.

payload   Content of a packet minus the headers; useful information.

PBX   Private Branch Exchange, telephone switch on customer premises.

PCI   Peripheral Component Interconnect, Intel's bus for personal computers; interface to accessory cards for sound, video, LAN connection, etc.

PCM   Pulse Code Modulation, how voice is encoded in a channel bank and other devices.

PDU   Protocol Data Unit, for a protocol level, the payload plus headers for that level.

PHY   PHYsical, the cable or fiber (or radio) transmission medium.

PKI   Public Key Infrastructure, certificate authority servers and software to sign and encrypt messages with a user's certificate.

PoE   Power over Ethernet, sends d.c. current over the LAN cable to run phone, video camera, or Wi-Fi access point.

POTS   Plain Old Telephone Service, the analog trunk.

power   Electrical measurement of the rate of work; of a CPU, the ability to perform a task in a given time, higher power translates to more work in less time.

PPP	Point-to-Point Protocol, L2 or data link protocol.
PPPoA	Point-to-Point Protocol over ATM.
PPPoE	Point-to-Point Protocol over Ethernet.
pps	Packets per second.
PR	Partial Reliability, setting in SCTP to limit number of retransmissions.
Proxy	SIP server that represents a phone to send and receive calls.
PRACK	PReliminary ACKnowledgment, sent before a final 200 OK message.
PRI	ISDN trunk and interface, one Data channel for signaling on 23 or 30 Bearer channels. 23B + D on a T-1 or 30B + D on an E-1.
private	Under sole control of a noncarrier; IP address that cannot appear on the Internet.
PS-ALI	Public Service–Automatic Location Information.
PSAP	Public Service Answering Point, receiver of 911 calls.
PSQM	Perceptual Speech Quality Measure.
PSTN	Public Switched Telephone Network.
PT	Payload Type, field in protocol header that identifies next encapsulated header.
PTR	PoinTeR, DNS record with cname or other idea for next search.
public	Open to use by all customers; IP address found on the Internet.
PVC	Permanent VC, configured in equipment, not under control of users.
QN	Quantizing Noise.
QoS	Quality of Service.
Q.sig	Common designation for Q.2921, a digital telephone signaling protocol for PBXs.
RAS	Registration, Admission, Status; procedures for end points in H.323 systems.
RAID	Redundant Array of Inexpensive (Independent) Disks.
RBE	Routed Bridge Encapsulation.
REFER	SIP message to direct a call to a different location or server.
reliable	Attribute of a protocol that ensures accurate delivery of packets.
request	Message from client to server that initiates a transaction
response	Message from server to client that answers a request
RFC	Request For Comment, original function of IETF document, now a final version.
RIR	Regional Internet Registrars, receive blocks of IP addresses from IANA, assigh them to ISPs.

RJ-45	Registered Jack #45, 8-contact modular connector used for 4-pair LAN cable.
ROHC	RObust Header Compression protocol.
ROI	Return On Investment.
rport	Reply port, needed to send packets back through a pinhole in a firewall.
RR	Resource Record, in DNS.
RRSIG	DNS record with a DNSsec signature to authenticate a listing.
RST	ReSeT, 1-bit indicator field in TCP header to reset the connection.
RSVP	ReSource reserVation Protocol, method for user device to request specific QoS on IP network.
RTCP	Real-time Transport Control Protocol.
RTP	Real-time Transport Protocol, L5, designed to facilitate "live" information streams.
RTP/AVP	RTP Profile for Audio and Video Conferences with Minimal Control, RFC 3551, a template for how to deploy for interoperability.
RTSP	Real-Time Streaming Protocol.
RU	Rack Unit, vertical height of 1.75 inches.
SACK	Selective ACK.
SAP	Service Announcement Protocol.
SASL	Simple Authentication and Security Layer, part of registration of user on IM server; RFC 6120.
SBC	Session Border Controller.
Scheme	URI designation that includes SIP, SIPS, TEL, and transport method.
SCLC	Synchronous Data Link Control, IBM's protocol for terminals on multidrop lines.
SCTP	Stream Control Transmission Protocol.
SCP	Service Control Point, feature server on SS7 neworrk.
SDES	Source DEScription.
SDLC	Synchronous Data Link Control, simple L2 protocol.
SDP	Session Description Protocol.
server	Receives and responds to SIP commands.
Session	Exchange of media stream between UAs.
SFTP	Secure FTP, encrypted.
SHA-1	Secure Hash Algorithm, encryption option within SSL and TLS.
SIGTRAN	SIGnaling TRANsmission, other formats carried over IP networks.
SIP	Session Initiation Protocol.

SIPS, SIP-TLS	SIP Secure, encrypted by TLS.
Skinny	Cisco's proprietary version of SIP.
SLA	Service Level Agreement.
SLC	Subscriber Loop Carrier, extensions of a CO switch located in pedestal cabinets.
SMB	Small or Medium Business.
SMS	Short Message Service, basis of Tweets and texting.
SN	Sequence Number, associated with a packet or (in TCP) an octet in a stream.
SNMP	Simple Network Management Protocol.
SOA	Start Of Authority, information about a DNS zone and administrator contact information.
SOAP	Simple Object Access Protocol, way to send business data in XML over HTTP.
socket	Combination of IP address and layer 4 port number.
SONET	Synchronous Optical NETwork, fiber optic architecture based on resilient rings.
space	Logical representation of a digital zero.
SP-ALI	Service Provider ALI, the one maintained by the LEC.
SPIT	SPam over Internet Telephony.
SQL	Structured Query Language.
sRTP	secure RTP, with encryption.
SRV	SeRVice, DNS record that contains addresses of servers for a domain.
SS7	Signaling System 7, based on packet switched data network among PSTN devices.
SSM	Source-Specific Multicast, only one sender on the tree.
SSRC	Synchronization SouRCe, primary intput to composite multicast stream (RTP).
ST	Straight Tip, one of several common connector types for optical fiber; resembles twist-lock BNC.
STP	Signal Transfer Point, packet switch in SS7.
STUN	Session Traversal Utilities for NAT.
SVC	Switched VC, set up on command/demand from user.
SYN	SYNchronize, 1-bit indicator field in TCP header; packet type that requests a new TCP connection.
T-1	Transmission level 1, 24 DS-0's in a 1.544 Mbit/s TDM channel on two UTPs.
T&S	Transmission and Switching.
TASI	Time Assigned Speech Interpolation.
TCO	Total Cost of Ownership, over the life of a device or project.
TCP	Transmission Control Protocol, connection-oriented L4 protocol above IP.

TD	Topology Descriptor.
TDM	Time Division Multiplexing, creates channels on a digital link.
TDoS	Telephony Denial of Service, attack to overwhelm a call center.
telco	Telephone company, any organization that transport voice for paying customers; LEC.
TEL	DNS record with an E.164 telephone number.
termination	Logical entity in MGW that sources/sinks media/control streams (H.248.1 §6.2).
TFTP	Trivial File Transfer Protocol, how IP phones get configurations and updates.
TLD	Top Level Domain: .com, .org, etc.
TLS	Transport Layer Security, SSL used for signaling and media connections.
TLV	Type-Length-Value, form of element or extension to protocol headers.
TN	Transport Number.
topology	Describes the flow of media among terminations in a MGW context.
TOS	Type Of Service, field in IP header.
TP0	Transmission Protocol zero, ISO terminology.
TPKT	ThroughPacket, multiplexed voice channels in one packet.
TR	Technical Recommendation.
Transaction	Exchange of messages to perform a task.
Transaction User	Agent or proxy that participates in message exchange.
trunk	Connectin or cable between customer and central office.
TSAP	Transport Service Access Point.
TSI	Time Slot Interchanger, the heart of a digital voice circuit switch.
TSN	Transmission Sequence Number.
TTL	Time To Live.
TURN	Traversal Using Relays around NAT.
TWG	Technical Working Group.
UA	User Agent, proxy for end point that finds paths to controllers and other end points
UAC	UA Client, issues requests.
UAS	UA Server, responds to requests.
UC	Unified Communications.
UDP	User Datagram Protocol, connectionless L4 protocol above IP.
UDPTL	UDP Transport Layer, used in fax machines.

UPF	Urgent Pointer Field, 1-bit indicator in TCP header.
UPS	Uninterruptable Power Supply, usually an inverter, battery, and charger to supply AC power during utility outages.
URI	Universal Resource Identifier.
URL	Universal Resource Locater.
UTF-8	Unicode Transformation Format, representation of alpha, numeric, and special characters; version 8 encodes HTML, SIP, etc.
UTP	Unshielded Twisted Pair, copper telephone wire without a metallic overwrap.
VAD	Voice Activity Detection, silence suppression.
VC	Virtual Circuit (Channel), path for packets or frames set up in forwarding tables among shared resources.
VJHC	Van Jacobson Header Compression.
vLAN	virtual LAN, ports and devices configured to recognize addresses in an additional Ethernet header field; separates users' traffic from each other.
VM	Virtual Machine, logical computer, many of which may run on the same hardware.
voice	Speech or audible tones audible to humans; POTS phones carry 300-3300 Hz.
VoIP	Voice over Internet Protocol.
VPC	Voice Positioning Center, application to track locations of phones for E911.
VPN	Virtual Private Network.
WAN	Wide Area Network.
Wi-Fi	Wireless LAN, 802.11.
WiMax	Worldwide interoperability for Microwave Access (Wireless to the Max), air interface standard.
X.25	packet switching standard and network service based on ITU recommendation.
XML	eXtensible Markup Language.
XMPP	eXtensible Messaging and Presence Protocol (RFC 6120), for IM.
ZBTSI	Zero Byte Time Slot Interchange, process to maintain 1's density on T-1 line.
ZRTP	Zphone Real-Time Protocol, method to exchange encryption keys in RTP.

12.2 REFERENCE DOCUMENTS

The most authoritative sources of technical details are the defining standards. Portions of this book draw on documents Copyright © by The Internet Society in the year originally published. All Rights Reserved by the IETF. As they say:

> IETF documents and translations of them may be copied and furnished to others, and derivative works that comment on or otherwise explain them or assist in their implementation may be prepared, copied, published, and distributed, in whole or in part, without restriction of any kind, provided that the above copyright notice and this paragraph are included on all such copies and derivative works.

Done.

IETF is only one of many standards bodies (Table 12.1) that cover the communications arena. Fortunately for the industry, these organization cooperate extensively and often publish an identical document bearing different numbers. Unfortunately, differences can arise.

More important, standards change—be sure you work from the latest versions. RFCs from the IETF don't change, but they are often superseded or amended by a new RFC with a higher number. Several sites publish annotated lists of IETF publications.

Of the references reviewed for this book was every page read? No, but they have all been skimmed. The author is happy to relieve the reader of that task.

12.2.1 RFCs

The IETF standardization process starts with a DRAFT document in a working group. It always carries an expiration date. If development continues, a draft may go through more than a dozen revisions, or it may expire and end there. When a draft stabilizes and vendors see enough value in it to implement at least two products that interoperate, the document may be published as an RFC. Those RFCs that see near-universal adoption and wide deployment may be designated an Internet standard, with another number (see Table 12.2). Of the more than 6000 RFCs published at this time, 68 are designated Full Standards (listed at http://www.apps.ietf.org/rfc/stdlist.html). Many more are "Standards Track" and widely implemented. IETF standards may become national and international standards when processed through ANSI and ISO.

RFCs are not revised, but obsoleted or updated by later RFCs on the same topic. There are errata files for some; check for corrections if an RFC is important to you.

In 2011 there were more than 180 IETF specifications for SIP. What happened to the others? Most didn't make this list because they deal with details outside the range needed to understand SIP and VoIP. Others are "informative"

TABLE 12.1   Organizations for communications standards

Institute of Electrical and Electronic Engineers
www.IEEE.org
Publishes standards for LAN's, Wi-Fi, and related materials.
Committee 802 usually offers documents more than 2 years old for free download; newer versions may be purchased.

International Telecommunications Union
www.ITU.int
The Telecommunications division publishes Recommendations for voice and data communications. A Radio division works similarly to standardize RF spectrum. The bulk of Recommendations are free downloads.

Internet Engineering Task Force
www.ietf.org
Check here for possible replacements, updates, or corrections of RFCs. All RFCs and published drafts are free to download. Some working documents are limited to committee members. IETF is an ANSI member.

International Standards Organization
www.iso.org
Sells paper copies and downloads (typically priced between 100 and 200 Swiss francs).
Publications are also available for purchase through ANSI, the US member of ISO.

American National Standards Institute
www.ansi.org
Member of ISO and IEC; represents the US. Accredits members like IETF and ATIS who develop consensus standards.

European Telecommunications Standards Institute
www.etsi.org
Free downloads of standards for registered users. No charge to register.

Internet Assigned Numbers Authority
http://www.iana.org
Operated by Internet Corporation for Assigned Names and Numbers: www.icann.org.
Assignments of well-known ports, IP ranges, etc., are on the website. Free.

Alliance for Telecommunications Industry Solutions
www.ATIS.org
Member of ANSI. Focuses on needs of carriers and their vendors in developing next-generation IP infrastructure and services.

TABLE 12.2   Key RFCs/Internet standards for VoIP/UC

RFC/ STD	Name	Length, Pages
768/6	UDP	3
793/7	TCP	85
1006	ISO Transport Service on top of the TCP Version: 3	18
2616	Hypertext Transfer Protocol—HTTP/1.1	176
2782	A DNS RR for specifying the location of services (DNS SRV)	12
3016	RTP Payload Format for MPEG-4 Audio/Visual Streams	21

(Continued)

TABLE 12.2   Key RFCs/Internet standards for VoIP/UC (*Continued*)

RFC/ STD	Name	Length, Pages
3261	SIP: Session Initiation Protocol	269
3262	Reliability of Provisional Responses in the Session Initiation Protocol (SIP)	14
3263	Session Initiation Protocol (SIP): Locating SIP Servers	17
3264	An Offer/Answer Model with the Session Description Protocol (SDP)	25
3265	Session Initiation Protocol (SIP)–Specific Event Notification	38
3289	Real-time Transport Protocol (RTP) Payload for Comfort Noise (CN)	8
3311	The Session Initiation Protocol (SIP) UPDATE Method	13
3312	Integration of Resource Management and Session Initiation Protocol (SIP)	30
3323	A Privacy Mechanism for the Session Initiation Protocol (SIP)	22
3327	Session Initiation Protocol (SIP) Extension Header Fieldfor Registering Non-Adjacent Contacts	17
3362	Real-time Facsimile (T.38)—image/t38 MIME Sub-type Registration	5
3389	Real-time Transport Protocol (RTP) Payload for Comfort Noise (CN)	8
3428	Session Initiation Protocol (SIP) Extension for Instant Messaging	15
3435	Media Gateway Control Protocol (MGCP) Version 1.0	210
3515	The Session Initiation Protocol (SIP) Refer Method	23
3525	Gateway Control Protocol Version 1	213
3550/64	RTP: A Transport Protocol for Real-Time Applications	104
3551/65	RTP Profile for Audio and Video Conferences with Minimal Control	44
3581	An Extension to the Session Initiation Protocol (SIP) for Symmetric Response Routing	13
3605	Real Time Control Protocol (RTCP) attribute in Session Description Protocol (SDP)	8
3665	SIP Basic Call Flow Examples	94
3711	The Secure Real-time Transport Protocol (SRTP)	56
3761	The E.164 to Uniform Resource Identifiers (URI) Dynamic Delegation Discovery System (DDDS) Application (ENUM)	18
3764	EnumService Registration for Session Initiation Protocol (SIP) Addresses-of-Record	8
3840	Indicating User Agent Capabilities in the Session Initiation Protocol	36
4032	Update to the Session Initiation Protocol (SIP) Preconditions Framework	10
4168	The Stream Control Transmission Protocol (SCTP) as a Transport for the Session Initiation Protocol (SIP)	10

TABLE 12.2 (*Continued*)

RFC/ STD	Name	Length, Pages
4117	Transcoding Services Invocation in the Session Initiation Protocol (SIP) Using Third Party Call Control	19
4566	SDP: Session Description Protocol	49
4567	Key Management Extensions for Session Description Protocol (SDP) and Real Time Streaming Protocol (RTSP)	30
4568	Session Description Protocol Security Descriptions for Media Streams	44
4612	Real-time Facsimile (T.38)—audio /t38	8
4733	RTP Payload for DTMF Digits, Telephony Tones, and Telephony Signals	49
4734	Definition of Events for Modem, Fax, and Text Telephony Signals	44
4787	NAT Behavioral Requirements for Unicast UDP	67
4856	Media Type Registration of Payload Formats in the RTP Profile for Audio and Video Conferences	29
4960	Stream Control Transmission Protocol	152
4961	Symmetric RTP/RTP Control Protocol (RTCP)	6
4975	The Message Session Relay Protocol (MSRP)	60
5031	A Uniform Resource Name (URN) for Emergency and Other Well-Known Services	15
5128	State of Peer-to-Peer (P2P) Communication across Network Address Translators (NATs)	32
5234/68	Augmented BNF for Syntax Specifications	15
5245	Interactive Connectivity Establishment (ICE): A Protocol for Network Address Translator (NAT) Traversal for Offer/ Answer Protocols	117
5322	Internet Message Format	57
5359	Session Initiation Protocol Service Examples (Centrex Features)	170
5389	Session Traversal Utilities for NAT (STUN)	50
5393	Addressing an Amplification Vulnerabilityin Session Initiation Protocol (SIP) Forking Proxies	20
5411	A Hitchhiker's Guide to the Session Initiation Protocol (SIP)	39
5506	Support for Reduced-Size Real-Time Transport Control Protocol (RTCP): Opportunities and Consequences	17
5589	Session Initiation Protocol (SIP) Call Control—Transfer	58
5626	Managing Client-Initiated Connections in the Session Initiation Protocol (SIP)	50
5630	The Use of the SIPS URI Scheme in the Session Initiation Protocol (SIP)	56
5761	Multiplexing RTP Data and Control Packets on a Single Port	13
5764	Datagram Transport Layer Security (DTLS) Extension to Establish Keysfor the Secure Real-time Transport Protocol (SRTP)	26

(*Continued*)

TABLE 12.2 Key RFCs/Internet standards for VoIP/UC (*Continued*)

RFC/ STD	Name	Length, Pages
5766	Traversal Using Relays around NAT (TURN): Relay Extensions to Session Traversal Utilities for NAT (STUN)	67
5876	Updates to Asserted Identity in the Session Initiation Protocol (SIP)	11
5923	Connection Reuse in the Session Initiation Protocol (SIP)	19
5928	State of Peer-to-Peer (P2P) Communication across Network Address Translators (NATs)	12
6026	Correct Transaction Handling for 2xx Responses to Session Initiation Protocol (SIP) INVITE Requests	20
6086	Session Initiation Protocol (SIP) INFO Method and Package Framework	36
6120	Extensible Messaging and Presence Protocol (XMPP): Core	211
6121	Extensible Messaging and Presence Protocol (XMPP): Instant Messaging and Presence	114
6122	Extensible Messaging and Presence Protocol (XMPP): Address Format	23
6140	Registration for Multiple Phone Numbers in the Session Initiation Protocol (SIP)	45
6141	Re-INVITE and Target-Refresh Request Handling in the Session Initiation Protocol (SIP)	26
6157	IPv6 Transition in the Session Initiation Protocol (SIP)	15
6189	ZRTP: Media Path Key Agreement for Unicast Secure RTP	115

(useful or interesting, but not on a standards track) rather than "normative" (requirements). Many RFCs describe things that never became popular or are obsolete.

12.2.2 ITU Recommendations

PSTN carriers used the recommendations of the ITU to guide vendors into make compatible and interoperable equipment (see Table 12.3). They also agreed on services, protocols, and interfaces that were adopted by carriers and often given the force of law in many countries.

12.2.3 Other Sources

In addition the documents in Table 12.4, the websites of hardware vendors and carriers supplied dozens of pieces of information on data sheets, white papers, user manuals, and reports.

TABLE 12.3 **ITU recommendations related to VoIP/UC**

Number	Name	Length
G.711	Pulse code modulation (PCM) of voice frequencies	10
H.225.0	Call signaling protocols and media stream packetization for packet-based multimedia communication systems	196
H.242	System for establishing communication between audiovisual terminals using digital channels up to 2 Mbit/s	100
H.245	Control protocol for multimedia communication	346
H.248.1	Gateway control protocol: version 3 (09/2005)	205
H.323	Packet-based multimedia communications systems, May 2003	320
H.460.18	Traversal of H.323 signaling across network address translators and firewalls	20
H.460.19	Traversal of H.323 media across networkaddress translators and firewalls	28
T.4	Standardization of group 3 facsimile terminals for document transmission	78
T.30	Procedures for document facsimile transmission in the general switched telephone network	322
T.38	Procedures for real-time group 3 facsimile communication over IP networks [under revision in 2011]	31
T.38 Amendment 1	Procedures for real-time group 3 facsimile communication over IP networks (Annex B)	15
T.120	Data protocols for multimedia conferencing	48
T.123	Network-specific data protocol stacks for multimedia conferencing	74
T.125	Multipoint communication service protocol specification	145

TABLE 12.4   **Other reference documents**

Number	Description	Size
	W. Flanagan, *Guide to T-1 Networking*, 5th Edition, Telecom Books, 1997	340
	William Flanagan, *Voice Over Frame Relay*, Flatiron Publishing, 1997	220
FRF.11	Voice Over Frame Relay Implementation Agreement	55
TWG-2	SIPconnect Technical Recommendation Version 1.1	45
LTRT-12802	SIP User's Manual, Version 6 (AudioCodes)	750
IEEE 802.3	Part 3: Carrier sense multiple access with Collision Detection (CSMA/CD) Access Method and Physical Layer Specifications [Ethernet]	671

12.3 MESSAGE AND ERROR CODES

TABLE 12.5   **SIP status responses**

3-Digit Code	Class	Explanation
1xx	Provisional	Request received, continuing to process the request
2xx	Success	Action was successfully received, understood, and accepted
3xx	Redirection	Further action needed to complete the request
4xx	Client error	Request contains bad syntax or can't be fulfilled at this server
5xx	Server error	Server failed to fulfill an apparently valid request
6xx	Global failure	Request can't be fulfilled anywhere

Source: From RFC 3261.

TABLE 12.6   Code numbers in UC signaling messages

Number	Meaning	Protocol	reQuest resPonse
100	Trying		
180	Ringing	SIP	
183	Session progress	SIP	
200	OK	SIP, MSRP	P
3xx	[See Table 12.7]	Warning header	
300	Multiple choices	SIP	
301	Moved permanently	SIP	
302	Moved temporarily	SIP	
400	Unintelligible request	MSRP	
403	Forbidden	SIP, MSRP	
404	Not found		
408	Downstream transaction timed out	MSRP	
413	Stop sending message	MSRP	
415	Media type not understood	MSRP	
416	Unsupported URI scheme		
420	Bad extension	SIP	
421	Extension required	SIP	
423	Parameter out of bounds	MSRP	
430	Flow failed	RFC5626	P
431	No terminationID matched a wildcard	H.248.1	P
439	First hop lacks outbound support	RFC5626	
481	No such session	MSRP	
488	Not acceptable here	SIP	P
501	Method not understood	MSRP	
506	Session ID used elsewhere	MSRP	
580	Precondition failure	IMS	
9xx	Service changes (rebooted, restored, MGW failure, etc.)	H.248.1, p.168	P

TABLE 12.7   SIP warning header codes

Code	Display Text	Meaning
300	Incompatible network protocol	One or more network protocols contained in the session description are not available.
301	Incompatible network address formats	One or more network address formats contained in the session description are not available.
302	Incompatible transport protocol	One or more transport protocols described in the session description are not available.
303	Incompatible bandwidth units	One or more bandwidth measurement units contained in the session description not understood.
304	Media type not available	One or more media types contained in the session description are not available.
305	Incompatible media format	One or more media formats contained in the session description are not available.
306	Attribute not understood	One or more of the media attributes in the session description are not supported.
307	Session description parameter not understood	A parameter other than those listed above was not understood.
330	Multicast not available	The site where the user is located does not support multicast.
331	Unicast not available	The site where the user is located does not support unicast communication (usually due to the presence of a firewall).
370	Insufficient bandwidth	The bandwidth specified in the session description or defined by the media exceeds that known to be available.
399	Miscellaneous warning	Can include arbitrary information to be presented to a human user or logged. A system receiving this warning MUST NOT take any automated action.

Note: Warn-Code values are registered with IANA.

INDEX

*VoIP and Unified Communications: Internet Telephony and the
Future Voice Network*, First Edition. William A. Flanagan
© 2012 John Wiley & Sons, Inc. Published 2012 by John Wiley & Sons, Inc.

277